Business, Organized Labour and Climate Policy

Business, Organized Labour and Climate Policy

Forging a Role at the Negotiating Table

Peter J. Glynn

Bond University, Australia

Timothy Cadman

Institute for Ethics, Governance and Law, Griffith University, Australia

Tek Narayan Maraseni

University of Southern Queensland, Australia

Cheltenham, UK • Northampton, MA, USA

Published by
Edward Elgar Publishing Limited
The Lypiatts
15 Lansdown Road
Cheltenham
Glos GL50 2JA
UK

Edward Elgar Publishing, Inc.
William Pratt House
9 Dewey Court
Northampton
Massachusetts 01060
USA

A catalogue record for this book
is available from the British Library

Library of Congress Control Number: 2016959920

This book is available electronically in the **Elgar**online
Social and Political Science subject collection
DOI 10.4337/9781786430120

ISBN 978 1 78643 011 3 (cased)
ISBN 978 1 78643 012 0 (eBook)

Typeset by Servis Filmsetting Ltd, Stockport, Cheshire
Printed and bound by CPI Group (UK) Ltd, Croydon, CR0 4YY

Contents

About the authors

Peter J. Glynn PhD (Bond University, Queensland), CPA, FCIS is an honorary adjunct Research Fellow at Bond University. His professional background is in industry association management, with specific expertise in labour relations and labour market management. Prior to commencing doctoral studies he was engaged by the International Labour Organization and the International Organisation of Employers in the development of policies and tools to guide the implementation of labour market strategies to manage the impacts of climate change policies. His doctoral research addressed the question of how industry (employers' organizations and trade unions) can ensure the impacts on employment and the workplace are considered in climate policy.

Timothy Cadman BA (Hons) MA (Cantab) PhD (Tasmania) is a Research Fellow in the Institute for Ethics, Governance and Law at Griffith University, Queensland, Australia. He is also a Senior Research Fellow in the international Earth Systems Governance Project and an Adjunct Lecturer and Member of the Australian Centre for Sustainable Business and Development at the University of Southern Queensland. He specializes in the governance of sustainable development, climate change, natural resource management including forestry, and responsible investment.

Tek Narayan Maraseni BSc (Tribhuwan) MSc (Asian Institute of Technology, Thailand) PhD (University of Southern Queensland) has over 20 years work experience in the areas of climate change adaptation, mitigation and greenhouse gas emissions accounting/modelling research in different countries including Nepal, Thailand, Australia and China. He has produced over 100 publications including two books in the last 10 years. His research work has been recognized through several national and international awards, grants and fellowships.

Preface

This book emerges from a discussion with Dr Tim Cadman of Griffith University over a glass of wine on the eve of the negotiations prior to the Paris COP at the UNFCCC meetings in Bonn in June 2015. Our collaboration for this book arises from our mutual concern about the gap between the emission reduction targets of the proposed Paris agreement and the reduction commitments of the Parties. We questioned the policy development process and debated the value of adding a theoretical framework to guide the negotiating process.

My interest in the international climate change negotiations and domestic climate change policy was sparked by a research project for the International Organisation of Employers in Geneva in 2008 as to whether climate change was an issue that required the attention of employers' organizations. The project found that climate change policy did impact the labour market, that labour market planning was not generally a part of policy and nor was it being prioritized by employers' organizations and trade unions. The journey since has sought to better understand how to effectively influence the policy development process and facilitate the transition to a low carbon workplace.

The climate change policy development process is highly politically charged and impacted by international competitive relationships and unique domestic economic, social and cultural priorities. In 2008, climate change was still very much regarded as an environmental issue and, although the realities for economies and societies are now accepted, the actions required for effective mitigation and adaptation are still insufficient to meet the challenge. Accepting the complexities the competing priorities impose on the policy development process and exploring what can be done to better guide that process, Dr Cadman and I reverted to our initial debate – could a theoretical framework add value to the negotiating process? We adopted ecological modernization as a suitable theoretical framework and then undertook the task to extend it for use as a guide in the policy development process. We had been challenged by the contentions of EM theorists that EM was increasingly in use in environmental policy analysis, yet the instances referenced were the occasions when policy aligned with the language or principles of EM, it was not presented as an influence over the decision or choices of the policymakers. In our view, EM had the potential to be an explicit guide in that process. Our work has presented EM in a form that can be used as a guide by policymakers in the achievement of their climate change objectives and helps explain and understand how the competing priorities mitigate policy impact.

We hope this book has provided some answers to questions about the impact of climate change on the labour market and the effectiveness of employers' organizations and trade unions in the governance of policymaking. We offer it as a contribution to updating ecological modernization for the contemporary policy environment. Hopefully the book can be a resource to which policymakers and practitioners may refer to better understand the issues and influences that are impacting the achievement of their climate change policy objectives, and that the gap between the commitment and ambition can be bridged.

Acknowledgements

I have been fortunate and privileged to be guided in my academic endeavours by my wife Dr Krishna Bose and for this book by Dr Tim Cadman. Krishna encouraged me to undertake the challenge of an academic pursuit after a career in industry which, as she had foreshadowed, has created opportunities for professional development at levels and in fields I had never anticipated. She has been my companion and publicist, allowed me the latitude to pursue the selfish tasks of academic research and the work for this book, and has maintained the family relationships that would otherwise have suffered in the pursuit of my personal goal.

I am of course indebted to Tim Cadman, a kind and generous taskmaster who accessed his contacts in the publishing world and offered patient editorial suggestions and control that has meant we are delivering an academic text that is also relevant to policymakers and practitioners. The statistical analysis by Tim's colleague Tek Maraseni was also invaluable for the survey analysis.

The experience and opportunity afforded by Bond University, which provided the platform for research and access to the tools necessary for my academic research, has been invaluable. Professor Taplin guided me through the process, allowing me the scope to draw fully on my experiences from industry and deliver a product acceptable to the academic community. I am grateful for her contribution and the management of our relationship.

The principal editor for this book is my daughter Belinda. I am immensely proud of her. Belinda has an established reputation in academic editing; she is very thorough, with the ability to see the logic (or the lack of logic) in a text. She has also very patiently nursed her father through the process to deliver his first book. The editorial effort was complimented by Ronita Bhattacharya, whose application and legal discipline ensured rigour was maintained in the recording of the many sources accessed to inform the research.

We acknowledge the publisher and in particular Alex Pettifer for his patience over the months that have passed and for accommodating the unforeseen issues that impacted our delivery schedule.

We also thank the employers' organizations, trade unions and other civil society who participated in this research.

Friends have also been very supportive, asking and encouraging. My other children, Paula and Jennifer, have helped make the journey a pleasure, always there to share the high and the lows and showing me that they are proud of what I have chosen to do.

In acknowledging these contributions and assistance, we place on this record that any mistakes in the text are our own.

Abbreviations

ACCI	Australian Chamber of Commerce and Industry
ACEI	Alliance for a Competitive European Industry
ACTU	Australian Council of Trade Unions
AIOE	All India Organisation of Employers
AITUC	All India Trade Union Congress
ASEAN	Association of South East Asian Nations
BDA	Confederation of German Employers' Associations
BDI	Bundesverband Deutschen Industries
BINGO	Business and Industry Non-governmental Organization
BIS UK	Department of Business and Industry
BusinessEurope	Confederation of European Business
CBDR	Common but Differentiated Responsibilities
CBI	Confederation of British Industry
CCA UK	Climate Change Agreement
CCC UK	Committee on Climate Change
CCL UK	Climate Change Levy
CCS	Carbon Capture and Storage
CDM	Clean Development Mechanism
CDP	Carbon Disclosure Project
CFTUI	Confederation of Free Trade Unions of India
CGT	French Confédération Générale du Travail
CIE	Council of Indian Employers
CII	Confederation of Indian Industry
CII-GBC	Confederation of Indian Industry Green Business Centre
CLC	Canadian Labour Congress
COP UNFCCC	Conference of the Parties
COTU	Central Organization of Trade Unions of Kenya
CSO	Civil Society Organization
DECC UK	Department of Energy and Climate Change
DGB	Deutscher Gerwerkschaftsbund Bundesvorstand
DRRM	Disaster Risk Reduction and Management
EBP	Evidence-based Policy

EC	European Commission
ECA UK	Electrical Contractors Association
ECO UK	Energy Company Obligation
EESC	European Economic and Social Committee
EFI	Employers' Federation of India
EM	Ecological Modernization
ENGO	Environmental Non-governmental Organization
EPI	Environmental Performance Index
ETS	Emissions Trading System/Scheme
ETUC	European Trade Union Confederation
ETUI	European Trade Union Institute
EU	European Union
FICCI	Federation of Indian Chambers of Commerce and Industry
FKE	Federation of Kenya Employers
FoE	Friends of the Earth
GDFC	The Green Deal Finance Company
GGKP	Green Growth Knowledge Platform
GHG	Greenhouse Gas
ICC	International Chamber of Commerce
ICFTU	International Confederation of Free Trade Unions
ILO	International Labour Organization
IMCCC	Singapore Inter-ministerial Committee on Climate Change
INDC	Intended Nationally Determined Contribution
INTUC	Indian National Trade Union Congress
IOE	International Organisation of Employers
IPCC	Intergovernmental Panel on Climate Change
ITUC	International Trade Union Confederation
KNCCI	Kenya National Chamber of Commerce and Industry
KP	Kyoto Protocol
LPAA	Lima Paris Action Agenda
LULUCF	Land Use, Land Use Change and Forestry
MEDEF	Mouvement des Enterprises de France
MENR	Ministry of Environment and Natural Resources
MNCs	Multinational Corporations
NAFTA	North African Free Trade Agreement
NATO	North Atlantic Treaty Organization
NCCS	Singapore National Climate Change Strategy
NGOs	Non-governmental Organizations
NSA	Non-state Actors

OECD	Organisation for Economic Co-operation and Development
SBF	Singapore Business Federation
SCOPE	Standing Conference of Public Enterprises
SMEC	SME Committee
SMEs	Small and Medium Enterprises
SNEF	Singapore National Employers Federation
SNTUC	Singapore National Trades Union Congress
TUC UK	Trades Union Congress
TUNGO	Trade Union Non-governmental Organization
UK	United Kingdom
UNCED	UN Conference on Environment and Development
UNCSD	United Nations Conference on Sustainable Development
UNEP	United Nations Environment Programme
UNFCCC	United Nations Framework Convention on Climate Change
UNFCCC APA	United Nations Framework Convention on Climate Change Ad Hoc Working Group on the Paris Agreement
UNFCCC AWG	United Nations Framework Convention on Climate Change Adaptation Working Group
UNFCCC SBI	United Nations Framework Convention on Climate Change Subsidiary Body for Implementation
UNHLPF	UN High Level Political Forum on Sustainable Development
WCED	World Commission on Environment and Development
WCL	World Confederation of Labour
WHO	World Health Organization
WIPO	World Intellectual Property Organization
WTO	World Trade Organization
WWF	World Wide Fund for Nature

Introduction: business and labour in climate policy

BACKGROUND

This book investigates the role of employers' organizations and trade unions in the development of climate change policy. It discusses the role of civil society actors including employers' organizations and trade unions in the policy development process, explores whether labour market considerations should be an element of climate change policy and applies ecological modernization and institutional governance theory to guide the consideration of these issues. Exploring the relationship between these practices, climate change policy and the labour market provides insights into the requirements of all stakeholders and informs the public policy process.

The authors are motivated by the belief that there is a need to inform policymakers on the labour market impacts and, equally, labour market practitioners of the impact that climate change policy will have on their sphere of interest. The objective of climate change policy is to reduce greenhouse gas (GHG) emissions, which requires change to patterns of production and consumption that has direct implications for the labour market. Public policy must take into account the labour market impacts of climate change or it may result in labour and skill shortages and displacement; these are issues that create barriers to the efficient conduct of business and the effectiveness of climate change mitigation and adaptation programmes. These issues can be avoided with proper planning.

The link between climate change and the labour market was formally recognized when the Heads of State and Government adopted the recommendations of the United Nations Framework Convention on Climate Change (UNFCCC) Adaptation Working Group (AWG) that climate change agreements provide for a just transition and decent work (UNFCCC AWG LCA, 2009). A just transition is defined as the recognition of workers' rights, decent work, social protection and social dialogue (Worldwatch Institute, 2008). The four tenets of decent work as articulated by the International Labour Organization (ILO) are creating good jobs, guaranteeing respect for workers and the recognition of their rights, extending social protection and promoting social dialogue (ILO, 2011a).

Government policies aimed at managing GHG emissions will require business to change its product and service delivery arrangements, which in turn means labour requirements will also change. With the increasing incidence of events attributed to climate change accelerating the timeframes and the radical changes required to address climate change, intervention in the labour market is necessary if labour is to be available in the numbers and with the skills required and if workers are to be afforded a just transition with rights and benefits as they transition to the new low carbon workplace.

While current international climate change commitments provide for the protection of workers' rights through the requirements for decent work and a just transition, they fall short of addressing the requirement for labour market reform. For labour market practitioners, this is deemed poor planning at best as disruption in the labour market is a certainty and, without addressing this requirement for reform, the chances of achieving ecological targets are diminished.

As a key actor in the regulation of labour markets and the development and enforcement of climate change policy, the state must also be considered in this discussion. Additional pressures on the state arising from the aftershocks of the global financial crisis, the discovery of new stocks of fossil fuels easing concerns about energy security and the growing ambition gap in international climate agreements are constant tests of the validity of the theoretical models that influence decision-making and public policy. The institutions involved with and responsible for policy development are being required to act quickly, yet the policy processes in which they are caught up often take years, if not decades, to negotiate outcomes acceptable to all parties.

The emergence of labour issues in formal climate agreements reflected the growing acceptance that climate change has impacts across the broader economy and society, and research published in 2007 and 2008 was instrumental in creating awareness of the employment and workplace impacts of climate change policy.

The Worldwatch Institute's (2008) report for the ILO encapsulates the findings that are common across research on the subject: due to the impact of climate change on public policy and the economy, there will be a consequential impact on the labour market and some jobs will be lost, some jobs will be created and some jobs will change. On balance, there will be a modest net growth in employment and all sectors of industry will be affected. It is contended by the ILO that governments must have policy to manage these changes and that social protection systems need to be in place to afford workers a just transition. The 2012 research by the International Labour Organizations' (ILO) International Institute of Labour Studies (ILO, 2012), updating the Worldwatch Institute (2008) Report reiterates the earlier findings, adding that outcomes for employment and incomes are largely determined by the policy instruments and the institutions (that is, the governance systems) that implement them, rather than being an inherent part of a shift to a greener economy.

This book is written at a time of increasingly erratic and disastrous weather patterns attributed to climate change and increasing acceptance across society of the need for action. Attention is shifting to the non-government sector for inspiration and initiative to support the international agreements that have not generated actions sufficient to contain the growth in global warming. The collegiate nature of decision-making in the European Commission, which includes employers' organizations and trade unions through the European Economic and Social Committee and across business more generally, suggests the opportunity exists for these interests to make a contribution to the development of effective climate change policy and also accept a shared responsibility for the present state of affairs.

INTERNATIONAL AGREEMENTS

International agreements are a pervasive influence over domestic climate change policy, particularly in the wake the of 2015 Paris Conference of the Parties (COP) during which the Parties (the member states) agreed to adopt the Lima Paris Action Agenda and its four pillars of the legally binding agreement, voluntary emission reduction commitment, finance and the role for non-state actors (NSAs). The background to the international negotiations and agreements provides an important insight to the emergence of climate change as a concern to humanity, the commitment to tackle the problem and the process issues to which resolution must be found for the effective implementation of the agreements.

Following the 1972 UN Conference on the Human Environment, as concerns about global warming and the future of the environment escalated, in 1983 the UN General Assembly established the World Commission on Environment and Development (WCED) to propose long-term environmental strategies for achieving sustainable development. The Commission's report *Our Common Future* (WCED, 1987) strongly influenced the Heads of State and Government in the lead-up to the 1992 UN Conference on Environment and Development (UNCED). The Conference, which became known as the Earth Summit, produced the Agenda 21 Plans for Action (UNCED, 1992), a blueprint for action to achieve sustainable development worldwide. It also produced the Forest Principles, the Convention on Biological Diversity and the Framework Convention on Climate Change (UNCSD, 2012a). Within five years of the Earth Summit, the 1997 Conference of the Parties (COPs) to the UN Framework Convention on Climate Change (UNFCCC) had agreed to the Kyoto Protocol (KP), which embraced the agreement between developed and developing countries to work together to meet the climate change challenge and developed countries would commit to emission reduction targets (UNFCCC, 1997). The 2015 Paris Agreement succeeded the Kyoto Protocol and sees all countries commit to programmes of emissions reduction.

KP set targets on the ratifying signatories, who represented 37 industrialized countries and the European community, for reducing GHG emissions (UNFCCC, 1997). The Paris Agreement provided that ratifying countries act in accordance with their intended nationally determined contributions (INDCs).[1] Almost all countries have declared their intension to ratify the Agreement. Even so, there is little consistency in the mitigation policy and programme development of the nation-states globally because the circumstances of each country are different and, as the OECD and others agree, there is no one solution that can be applied to all.

The European Union (EU) was an active participant in the 1992 Earth Summit and a signatory to the subsequently ratified 1997 the Kyoto Protocol. The 1993 Treaty of Maastricht added the protection of the environment to the EU objectives of economic growth and social wellbeing (Europa, 2012a). Prior to the ratification of the Treaty of Maastricht, many countries within the EU had already moved to reduce their GHG emissions and to address concerns about energy security and the development of alternatives to fossil fuels (Syndex, 2011). In doing this, the EU and member states provide a working model of ecological governance that is comprehensive, mature and respects the autonomy of the state in determining domestic policy while operating within a regional framework. The EU and member countries provide a model of sustainability management and public policy development and as such are a rich source of valid data for research.

The attention afforded the economic and social issues alongside the climate imperative and the perceived needs of civil society has impacted the efficiency of the international climate change agreement negotiating process and efforts to meet the global warming targets. The 2011 Durban COP was criticized for its weak outcomes and slow progress and was generating a concern among stakeholders that the negotiators would be able to reach agreement on a successor to the Kyoto Protocol, concerns that have now been mitigated to some extent by adoption of the 2015 Paris Agreement and the commitments to further action.

The agreements from the Rio + 20 Conference (UNCSD, 2012a) and the UNFCCC COPs are major influences on the decisions of the state and the policy choices for climate change mitigation strategies. The agreements have as their objective sustainable development, which they describe as the delivery of environmental protection, economic growth and social wellbeing (UNCSD, 2012a; UNFCCC, 2011b). The Bali Action Plan (UNFCCC, 2007) prioritized the requirement that the economic and social consequences of adaptation to climate change be considered in negotiations. The Rio + 20 outcomes document *The Future We Want* (UNCSD, 2012a) contains these provisions as well as others addressing directly the issues of poverty eradication, the engagement of civil society, a green economy and institutional arrangements.

While comprehensive and well-intentioned, these requirements may have created a barrier to the effective implementation of the agreements. This is reflected in the growing concern among observers and interest groups that the climate agreements may not deliver on their target of containing the average increase in global temperature to less than 2 degrees, that civil society is becoming disenfranchised, that negotiators have so far failed to address the ambition gap and that there may be claims for compensation from developing countries and disputes over intellectual property rights (for example, Climate Action Network, 2012; Fisher, 2010; Eastwood, 2011). The outcomes from the 2015 Paris COP, while yet to be implemented, saw the Parties respond in a positive manner and provide hope that these concerns have been addressed.

The outcome agreements from the recent UN events including Rio + 20 and the UNFCCC COPs have been lauded as either positive demonstrations of multilateralism or criticized as insufficient to address the environmental concerns (for example, WTO, 2012; Climate Action Network, 2012; BusinessEurope, 2012). Time-consuming and unresolved issues from the UNFCCC COPs at Durban and Doha included intellectual property rights, finance and claims by developing nations for disaster compensation, issues which can only be described as incidental to the environmental objective (BusinessEurope, 2012). The 2010 Cancun Agreement (UNFCCC, 2010) and Rio + 20 outcome agreement added commitments to labour reform. The merits or otherwise of these additions are not at issue: at issue is whether international environmental and climate fora are the appropriate places for them to be negotiated and whether they encumber the ability to negotiate agreements that could address the ambition gap. The 2012 Doha COP heard arguments by developed countries that other fora such as the World Trade Organization (WTO) and World Intellectual

Property Organization (WIPO) would be better placed to resolve intellectual property issues (Climate Action Network, 2012). The same could be said of the ILO in respect to the emerging labour issues. It is noted that these issues have not been carried forward into the decisions from the Paris COP, which may bode well for efficiency of future COPs and the implementation strategies. A legally binding core agreement stripped of these and the other non-core issues that 'cluttered' the Cancun and later COPs may improve the efficiency of decision-making.

The Rio + 20 outcome agreement included in its work programme an employment and workplace dimension by providing for the promotion of a just transition, decent work and social protection (UNCSD, 2012a). The UNFCCC climate change agreements had added a similar commitment at its 2010 Cancun COP (UNFCCC, 2010). The emergence of labour issues in formal sustainability and climate agreements reflects the growing acceptance that climate change has impacts across the broader economy and society.

The labour matters introduced into the Rio + 20 agreement at Articles 147–157 are comprehensive and present a thorough checklist of rights and obligations. They are also under the authority of specialized international agencies and in many countries are already principles of national labour law and practice (ILO, 2011). The potential jurisdictional conflict is mitigated to an extent by use of terminology that has wide application and recognition in formal international and industrial texts. This could imply a UN intention to endorse existing law and practice rather than to create new benchmarks for performance, even though the Parties are testing their authority in other fora.

The Paris COP, held in December 2015, adopted the framework of the Lima Paris Action Agenda (LPAA) and its four pillars: the binding Paris Agreement, voluntary country contributions (INDCs), finance and an enhanced role for NSAs. The Paris Agreement introduces a management model to ensure transparency in the reporting of voluntary undertakings that will operate in concert with the binding Paris Agreement and a modified system for monitoring and evaluation (UNFCCC, 2015, paragraphs 85–104). These arrangements will be integral to the effectiveness of the LPAA and of relevance to the governance framework presented in this volume.

The negotiations in the period leading up to the 2015 Paris COP were tense and closely monitored. The UNFCCC regular mid-term meetings of the Parties to the Convention in June 2015 in Bonn was the penultimate negotiating session and, while the Parties expressed their commitment to a process publicly, this did not translate to progress on streamlining the negotiating text. The meeting was also coming to terms with the need to share responsibility with non-Party actors, a reflection of acceptance that the targets for global warming and GHG emission containment could not be achieved by the states acting alone. This is a material shift whereby the concepts of shared responsibility and shared decision-making were to be introduced into UNFCCC governance, a role beyond business as usual for Parties and the civil society organizations (CSOs).

The objective of the UNFCCC is to stabilize greenhouse gas concentrations at a level that would prevent dangerous anthropogenic (human-induced) interference in the climate system (United Nations, 1992). The 2010 Cancun COP resolved to commit to action that would contain warming to a maximum temperature rise of 2 degrees Celsius above pre-industrial levels and to consider lowering that maximum to 1.5 degrees in the near future (UNFCCC, 2010), a commitment reiterated at the 2015 Paris COP (UNFCCC, 2015). The IPCC 5th Assessment Report (IPCC, 2014), however, provides a sobering review, reporting that without additional efforts to reduce GHG emissions beyond those in place today, global emissions growth is expected to persist (Christ, 2014) putting the 2-degree target out of reach.

The negotiations for the agreement to succeed the Kyoto Protocol began in 2007 with the Bali Action Plan (UNFCCC, 2007), intended to be finalized for resolution at the 2009 Copenhagen COP. The Parties at Copenhagen did not adopt the negotiated text and subsequent COPs also failed to agree terms for the next generation treaty. The Parties, however, rose to the occasion of the deadline of the 2015 Paris COP delivered a strong outcome. Time will tell whether the outcome and the enthusiasm from the COP are sufficient to deliver the 2-degree global warming objective and to bridge the gap between the voluntary commitments of the INDCs and the emission reductions required.

In his address to the Plenary of the UNFCCC meetings of June 2015, the incoming COP President, Laurent Fabius, France's Foreign Minister, foreshadowed the intended structure of the outcomes from the twenty first COP and the eleventh Conference Meeting as Parties to the Protocol (CMP) (COP21/CMP 11) held in Paris 30 November–11 December 2015 (Fabius, 2015). The stated vision of the French government for Paris 'focuses on actions pre 2020, as well as afterwards' (p. 1). Its Action Agenda sought to trigger the growing engagement of NSAs and proposed dialogue during the COP with national governments, cities and businesses (UNFCCC LPAA, 2015), effectively harnessing the willingness of the major corporations to act (Cameron et al., 2015) and tap finance opportunities from the private sector (Business and Climate Summit, 2015). The dialogue was conducted as proposed and the announcements of their commitment and proposed actions post-COP by these non-Party stakeholders is testament to the vision and effectiveness of the strategy (Food and Beverage Leaders, 2015; Gillis and St. Fleur, 2015; Pianigiani, 2015).

Reaching out to the business community and others who, by their own actions can reduce GHG emissions and impact the effectiveness of adaptation and mitigation strategies, is a logical and mature rationale. It does, however, introduce other considerations, importantly the monitoring and evaluation process. It also has the effect of subordinating the preeminence of the agreement that will be binding on governments, being only one of the four pillars of action and then with the key role as a technical standard-setting authority for measurement and reporting. It can be said that the four-pillar model is a reflection of society's concerns for the issue and its resolve to act (and the inability of the state alone to deliver the targeted outcome), but this comes at the expense of the capacity to enforce binding limits and timeframes.

The 2015 Paris COP has formally adopted an agreement in two parts: the core agreement and decisions. The core agreement (the Paris Agreement) will be legally binding on all parties and will address the ambition, requirements for adaptation and mitigation, financing and implementation. The Decisions of the COP contain the detail that may be varied as required and appropriate. The further voluntary commitments by each Party to emission reduction will be through the INDC. The four pillars of the Action Agenda as described earlier are the legal agreement, ambitious INDCs, long-term finance and a commitment from NSAs. This is a model that could deliver strong and appropriate governance; however, the success of the model is contingent on bridging the outstanding gap between the aggregate of the INDCs and the emission reductions required for containing global warming to less than the 2-degree objective. The Decisions document is precise, clear and provides a detailed and thorough work plan for the Parties and the UNFCCC Secretariat. It does not make the promises of the previous agreements to do everything for everyone, such as the labour and trade issues already referred to above. The Agreement is more forgiving but is still focused on the climate issues.

The outcome from the 2015 Paris COP is widely agreed as a solid platform for managing the policy necessary to address the issues arising from climate change and to that end the Paris COP is described as the negotiating event and COP 22 in Morocco in November 2016 as deciding the implementation arrangements. The Paris COP and the attention it created around the problem of climate change has inspired a number of parallel initiatives by governments such as the Coalition of Carbon Pricing (made up of governments that are now implementing or are proposing emission trading schemes) and the Renewable Energy Initiative (made up of governments and other stakeholders).

In respect to a global climate change initiative, it is a good outcome, and workable. For the business community, it is also a good outcome: it provides some certainty for decision-making; it does not presume to impose regulatory requirements over business and avoids issues relating to market behaviour and market mechanisms. It imposes obligations on the state for recording and reporting, but otherwise all issues relevant to business and possible regulation are contained in the governments' already submitted intended nationally determined contributions (INDCs).

In sum, the 2015 Paris COP agreed a framework for the management of climate change that is less centralized and relies more on the voluntary initiatives of the Parties individually, the private sector through the greening of their own activities and those in their supply chains, the financing of adaptation and mitigation and the broader network of NSAs. This is a bottom-up model which is different to the top-down model of Kyoto, and civil society resources need to be redirected accordingly, a change that may benefit many CSOs and allow many to play to their strengths in domestic policy. The converse may also be the case; CSOs may become less relevant as the UNFCCC creates an interface for direct dialogue with the broader community of NSAs.

CASE STUDIES

The book is built around a study of employers' organizations and trade unions, providing a unique set of perspectives of key stakeholding organizations. It draws from the experiences in eight countries (the United Kingdom (UK), France, Germany, Australia, Canada, Singapore, India and Kenya) and the European Union, with a detailed analysis of the European Union and the UK. The profiles were developed from a literature search of the governments, employers' organizations and trade unions in the countries selected. A survey was conducted across a broader community of trade unions and employers' organizations to gauge business and labour perceptions about their participation in climate-related policymaking. The results of the survey are used to inform the broader analysis of the book.

The European Union provides an established model of sustainability management and related public policy development and as such is a rich source of valid data for investigation. It is for this reason that the EU and member states with mature developed economies were selected for the case studies. The European Union (EU) was an active participant in the 1992 Earth Summit and a signatory to the subsequently ratified 1997 Kyoto Protocol. The 1993 Treaty of Maastricht added the protection of the environment to the EU objectives of economic growth and social wellbeing. Prior to the ratification of the Treaty of Maastricht and because of concerns about energy security, many countries within the EU had already moved to reduce their GHG emissions and to develop alternatives to fossil fuels. The pursuit of this objective has led to the creation of an ecological governance model that respects the autonomy of the state in determining domestic policy while operating within a regional framework.

CENTRAL RESEARCH QUESTIONS AND MEANS OF ASSESSMENT

The book asks the following questions:

1. Is the labour market significantly impacted by climate change policy?
2. Are employers' organizations and trade unions important actors in the development and implementation of climate change policy?

This book is informed by the theoretical framework of ecological modernization (EM), which contends that the relationship between the nation-state, the economy and innovation, and civil society are integral in the achievement of environmental outcomes – a theory that is increasingly used to guide public policy development. An important element in ecological modernization is civil society, which is often directly or by inference a reference to environmental activists. The book builds on and adds to existing theory by exploring the significance of labour market issues in climate change policy and the role of employers' organizations and trade unions as a further element and examining whether EM theory holds. As articulated by Mol et al. (2009), EM theorists agree that further research is required and that 'little is known still on how, to what extent and how successfully environmental interests are included in all kind of economic, cultural and political practices' (p. 511).

The intention of this book is to contribute to the body of knowledge that informs the theory of EM. It discusses the validity of the theoretical framework of EM and the gaps that exist between the theoretical framework and international and domestic policies, and how EM can and should be introduced into future policy development, monitoring and evaluation.

In view of the fact that EM theory is primarily directed towards understanding the interactions between the state, economy, ecology, and society, it is particularly relevant to studies of the governance of climate change management (Bailey et al., 2011). Developments in theory over the past couple of decades view the idea of governance as a means of coordinating the processes and structures of governing that goes beyond traditional government/public administration models (Rhodes, 1997; Salomon, 2002). Governance, as opposed to government is now seen as a more relevant term for understanding the nature of non-state and state interactions in international policy, particularly when policymaking involves the market and society as well as the state (van Kersbergen and van Waarden, 2004; Kooiman, 2000), hence the strong conceptual linkages to EM theory.

In such a setting, institutions of policymaking are understood as placing more focus on structural and procedural aspects than models of governing based on command and control and a greater emphasis on the participation of stakeholders in decision-making (Pierre and Peters, 2000). In intergovernmental policy regimes such as the UNFCCC, the manner in which different actors behave towards each other determines how regimes are constructed and constituted and the subsequent quality of stakeholder interactions makes a significant contribution to governance effectiveness and institutional legitimacy overall (Figure I.1).

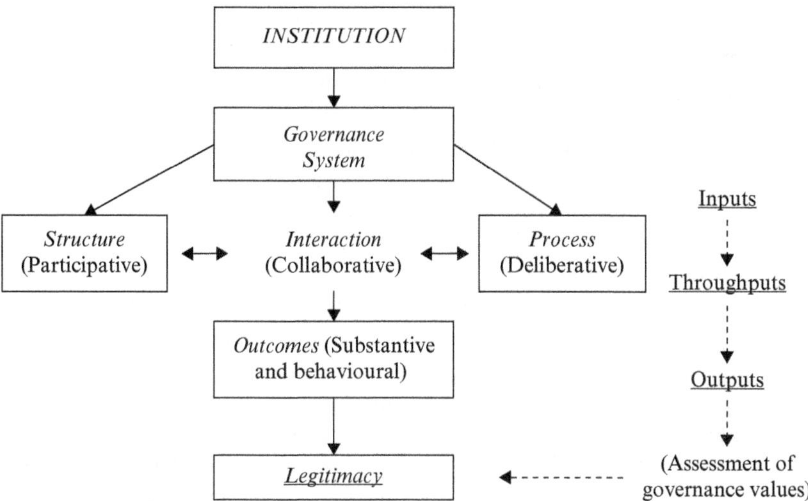

Source: Cadman (2011, p. 5, adapted; courtesy of Palgrave Macmillan).

Figure I.1 Analytical model for the evaluation of governance legitimacy

APPROACH AND METHODS OF EVALUATION

Ecological Modernization and the Case Studies

In the period between the 2010 Cancun COP and the 2015 Paris COP there was increasing concern among observers and interest groups that the climate agreements may not deliver their target of containing the increase in global temperatures. Civil society was becoming disenfranchised, negotiators could not agree measures to address the ambition gap, there were claims for compensation from developing countries and disputes over intellectual property rights (for example, Climate Action Network, 2012; Fisher, 2010; Eastwood, 2011). These events lead to the question as to whether an operationalized theoretical framework such as EM can guide the policy development process and policy choices towards effective climate change policy.

The negotiation of the international agreements is a complex and lengthy process that demonstrates the commitment of the states to the environment and, in particular, to addressing climate change; a commitment that it is expected will guide their approach to domestic policy development. The agreements are not framed by a theoretical model and an operationalized EM may best serve the states in the adaptation of the international and domestic policy development processes as states adapt to the economic and social changes necessary to reduce their GHG emissions. Each nation-state has unique political and cultural sensitivities that influence policy choices, which may also be a constraining factor over the implementation of other versions of EM.

To establish if significant gaps exist between the EM framework with the UNFCCC agreements and domestic climate change policy, a test was conducted in a sample of countries to inform investigation into how EM can be operationalized and to determine its relevance as a guide in the policy development process. This book overlays an EM template intended to identify differences between the elements that make up EM and the climate change policies of the chosen sample of climate agreements and signatories (UNFCCC itself, EU, UK, China and India) thereby highlighting if and where their policies diverge from the EM model (Chapter 2). The test found there were differences, for example, whereas an ideal objective of EM is the decoupling of economic development and environmental protection, the UNFCCC seeks to achieve the multiple objectives of containing global warming, promoting economic growth and social development; the European Union and UK have added energy security as an imperative, and in India and China, the institutions of the state, market and civil society are weak. Until the 2015 Paris COP, the UNFCCC agreements also exempted some of the world's major GHG emitters (China, India, Brazil and Indonesia) from legally binding commitments to reduce GHG emissions through the common but differentiated responsibilities (CBDR) provisions. While preliminary, the test suggests that the expectations of the contribution of the theoretical framework can only be to that common aspect of the defined objectives.

In essence, the theoretical framework helps evaluate the performance of climate policy using a set of measurable conditions. To that end and, as established in the experiment, an operationalized EM would serve to identify where there are gaps in the suite of policies and programmes that may inhibit the ability to achieve the targeted ecological outcome, recognizing there may be other factors that could also create such a barrier but are outside the scope of the theoretical framework.

The UNFCCC's Kyoto Protocol (UNFCCC, 1997) and the INDCs under the 2015 Paris Agreement provide for ratifying nations to commit to specific targets and timeframes concerning emission reductions and many member states, particularly in developed economies, have framed domestic policies with these targets as outcomes. An examination of the INDCs of the countries profiled in this study provides an opportunity to explore the ability to guide the policy development process in pursuit of the emission reduction target. In Table I.1 a qualitative scoring has been attached to the case studies, based on the relationship between the INDCs and the EM framework. The identified assessment parameters of the EM framework presented in this book are innovation and technology, the market, interventions of the state, the role of civil society, and ecological consciousness (see Chapter 2). While EM theorists generally speak of EM on a scale of weak to strong, it was considered that policymakers would be better assisted by a more definitive measure. To that end and with consideration of the Pathways to operationalize EM as discussed in Chapter 2 measures of low, medium and high were adopted in place of the weak to strong scale, and the values of 0, 5, and 10 were applied to low, medium and high (weak to strong) models of EM. The results are reflected in Table I.1.

In Table I.1, overlaying the EM template permits a focus on the core issues necessary to the achievement of the desired outcomes and identifies those extraneous issues that can be used to evaluate policy priorities and choices. The table demonstrates, for example, that for Kenya and India, policies are conditional on international technical and financial support, and in the case of India, institutions of the state are weak and the capacity to support the strategies is limited. Australia has not aggressively tackled the problem and Canada has a varied history. With the discovery of coal tar sands, Canada withdrew from the Kyoto Protocol, although the

Table I.1 EM template overlaid on INDCs

	Kenya	Australia	Singapore	Canada	India	UK	France	Germany	European Union
Market	5	5	10	5	5	10	10	10	10
Innovation	10	5	10	10	5*	10	10	10	10
State	10**	5	10	10	5**	10	10	10	10
Civil society	10	5	10	10	0	10	10	10	10
Ecological consciousness	10	5	10	5	5	10	10	10	10
	45	25	50	40	20	50	50	50	50

Notes:
* Government is actively pursuing the introduction of new technology and finance from other countries though the technology transfer arrangements under the UNFCCC agreements.
** Subject to international assistance.

country's recent change from a Conservative to Liberal government sees it revert to support for the international process.

Overall, the investigations conducted in this book find that while climate-related agreements and policies examined can be described as demonstrating elements of what some scholars refer to as 'strong' EM, multiple objectives introduce factors that mitigate against the successful application of the EM framework – and consequently the optimum policy choices to deliver the ecological outcomes desired. Chapter 1 explores these issues in more detail, presenting a new approach to evaluating policy effectiveness against EM theory, which moves away from previous notions of 'weak' or 'strong', and suggests a more nuanced approach instead.

Stakeholder Perspectives on the Quality and Legitimacy of Climate Governance

In addition to the application of EM theory to the case studies selected, the book also presents the results of a survey aimed at determining and evaluating the perspectives of civil society (such as employers' organizations and trade unions) on the governance arrangements for climate change policymaking at the intergovernmental, regional (EU), national, local and other levels. Specifically, survey participants were asked to rate their perspectives on how meaningful they thought their participation was in climate policy and how productive they thought the related policy deliberations were. Having established the linkages between EM theory and governance legitimacy in the discussions above, it is important to stress that consistent methods of evaluation are also required to determine whether the governance of policymaking sufficiently expresses the collaborative arrangements required to combat climate change effectively. Table I.2 contains a comprehensive set of governance values using a hierarchical framework of principles, criteria and indicators (PC&I), derived from a review of contemporary governance literature (Cadman, 2011).

PC&I represent a means for assessment of sustainability, including climate change, and were popularized as a consequence of the 1992 Rio Earth Summit and reflected in its foundational document *Agenda 21* (Rametsteiner et al., 2009). These sorts of framework arose to enable more consistent evaluation. This is achieved through the placement of each element in its appropriate location, from principles to criteria and from there to the appropriate indicator, to avoid overlap and duplication. A principle is an essential rule or value to be determined and is broken down into criteria, which for assessment purposes are further divided into indicators, also known as parameters, for measurement. Principles and criteria are ideational rather than directly measurable; hence assessment

Table I.2 Hierarchical framework of governance values

Principle	Criterion	Indicator
Meaningful participation	Organizational responsibility	Accountability
		Transparency
	Interest representation	Inclusiveness
		Resources
		Equality
Productive deliberation	Decision-making	Democracy
		Agreement
		Dispute settlement
	Implementation	Behavioural change
		Problem-solving
		Durability

Source: Cadman (2011, p. 17, adapted). Reproduced courtesy of Palgrave Macmillan.

itself occurs at the level of the indicator (Lammerts van Beuren and Blom, 1997, pp. 5–35).

As a principle, meaningful participation relates to the governance structures in which stakeholders are involved. It is comprised of two criteria: interest representation and organizational responsibility. Three indicators make up interest representation: inclusiveness, which is concerned with who participates in the given institution (delegated representatives of state or non-state organizations); equality, which relates to the balance of power between different interests; and resources, which refers to the capacities (such as technical, institutional and/or financial) participants have to draw on in ensuring their interests are represented. Organizational responsibility consists of two related indicators: accountability, addressing the extent to which participants can be called to account by others within the system and the general public; and transparency, referring to the degree to which participants' actions are visible or open to scrutiny by others in the system and the public. The principle of productive deliberation is concerned with the processes of governance and is comprised of two criteria: decision-making and implementation. Decision-making contains three indicators: democracy, which applies to the arrangements in place for selecting preferences; agreement, pertaining to the manner in which preferences are selected (such as consensus or voting); and dispute settlement, which is concerned with the mechanisms for dealing with disagreements and/or conflict. Implementation includes three indicators: behavioural change, that is, whether the decisions made lead to modified conduct – in this case, in relation to altering practices that lead to dangerous greenhouse gas (GHG) emissions; problem-solving, which concerns itself with the degree to which the initial predicament that led to institutional formation is successfully addressed – in this case in relation to the successful reduction of GHG emissions; and durability, encompassing the notions of adaptability, flexibility and longevity of the problem-solution. Such an approach makes a useful contribution to EM, as it allows for a comprehensive analysis of the institutional structures and processes underlying policymaking, which underpins the achievement of desired environmental outcomes.

KEY FINDINGS AND CHAPTER OUTLINE

Key Findings

The investigations in this book establish the effectiveness of public policy, and the role of stakeholders in structures and processes of climate change-related policymaking. Preliminary indications are that domestic labour market planning is essential in responding to climate change, yet labour market and climate change policies are not linked. Second, although employers' organizations and trade unions are influential as climate change activists within civil society, they are not effective advocates when it comes to substantively influencing climate change policy or attracting attention to the labour market impacts of climate change. The investigations also highlight the importance of understanding the governance of climate change policy and its contribution to monitoring and evaluating the role of NSAs such as business and labour in policymaking.

The findings of the book are intended to provide guidance on the role and effectiveness of business and labour in climate change policy. In the debate about climate change and the public policy intervention, the need and benefit of labour market planning is often overlooked. Labour market planning needs to be included in climate change policy. Strategies to manage GHG emissions are uniquely local and reflect many influences, such as the domestic economic, social, cultural and political situation of a country. However, all of these strategies have an impact on the labour market in addition to the normal and customary turnover of labour. Adjustments to labour planning by the state and industry need consideration if the demands for labour by industry are to be met and so that workers will be treated fairly during the transition to and in a new low carbon economy.

The book also offers an opportunity to explore the role of civil society. While EM theorists' references to civil society often directly or indirectly imply environmental activists, the book concentrates more specifically on how ecological modernization theory embraces the activities of the employers' organizations and trade unions, but also looks in passing at other interest groups formally recognized by the UNFCCC as civil society. Research by advisory bodies such as Eurofound (2011a), CEDEFOP (2010), OECD (2013b) and ILO (2012) finds that social partners, employers' organizations and trade unions have an important role to play in the development of climate change policy and the transition to a low carbon economy. The book also finds that the institutional framework of the EU confers authority on the social partners even though they are not necessarily the most capable or most representative organizations. In practice, the contribution of the social partners in the member states to the process of developing climate change policy reflects different levels of commitment. Their work programme does not always prioritize climate change but rather reflects the current issues of the day.

The book also establishes that what is a priority issue in one state may bear no resemblance to the work programme in another state that has a different economic profile and where the history, culture and tradition has cultivated different relationships. The UK study in particular demonstrates how culture, history and tradition can frame the statutory influences, attitudes and priorities of an organization. For example, with labour relations, in which the UK government has traditionally adopted a non-interventionist approach, negotiations occur at the level of the workplace and collective bargaining is voluntary. This means that the members' expectations for service by the Confederation of British Industry (CBI) and the Trades Union Congress (TUC) are different from the service delivery expectations of employers' organizations and trade union members in other EU member states where the labour market and the role of the social partners is more regulated. The EU member states are also at different stages of economic and institutional development, therefore the contributions by the social partners and the outcomes of social dialogue across the member states will vary. This is true also for countries outside the EU, such as Kenya and India.

The results of the governance survey are interesting. Overall, none of the levels of climate-related policymaking performed especially well in terms of perceptions of governance quality among survey respondents. There was a general preference for policymaking at the national level, notably amongst EU respondents. Employers' associations which participated in the survey did not view policymaking as being particularly transparent or accountable, but they were generally satisfied with its inclusiveness and equality. Trade unions, on the other hand, did not have such issues with accountability and transparency but were more concerned about issues of inclusiveness and equality. All respondents indicated that the level of resources afforded to them by policymaking institutions was very low. Additionally, respondents also generally perceived the problem-solving capacity of climate policymaking to be weak. Here it may be possible to determine a sense of frustration amongst respondents with both the structures and the processes of climate policymaking. While they may be able to forge a seat for themselves at the negotiating table to a certain degree, what they can achieve once they are there is questionable.

In view of the fact that EM is now over thirty years old, this book consequently suggests some modifications to EM theory in the light of current developments. Here, an understanding of the constraints placed on ecological outcomes by policy processes themselves, as well as more detailed governance assessments, focusing on the role of civil society actors, would help clarify EM theory. The model to operationalize EM theory presented here suggests that different contexts generate different policy frameworks and outcomes. The model also suggest that the achievement of outcomes is dependent on the policy choices generated as a consequence of distinctive political, institutional and cultural features, the national economic importance of specific sectors and, for example, the extent of the environmental impact of particular industries.

The book also stresses the importance of creating and maintaining political interactions that are transparent and democratic and ensuring that social movements are included in policymaking (Howes et al., 2010; Christoff, 1996). Within this scenario, the role for social partners (such as employers' organizations and trade unions) is as civil society actors whose opportunities are enhanced in the 'stronger' versions of EM. If climate policymaking is to afford civil society those opportunities, greater attention must be given to enhancing the interest and representation of NSAs (including enhanced institutional and technical support) as well as demonstrating greater levels of accountability and transparency. Without these, the implementation capacity of policymaking will remain constrained due to power asymmetries between state and NSAs – and climate change will remain an intractable issue.

Chapter Outline

Having introduced the subject matter, case studies and the approaches and methods relevant to understanding the role of business and organized labour in climate change policy, Chapter 1 discusses the understanding of and approach to ecological modernization used in the book. It examines EM in its historical and current contexts, the variations in the interpretation of the components of EM, and their interactions. It further explores the operationalization of EM from a theoretical concept to a practical framework that can guide public policy, and presents a new approach for doing so. The chapter reflects on the potential for conflict between the competing priorities of EM's theoretical framework and the policy constraints arising from the imperative to satisfy multiple objectives. Chapter 2 looks at the labour market impacts of climate change and the role of employers' organizations, trade unions and civil society on policymaking in different countries and policymaking forums. Chapters 3–6 comprise the case studies that inform the research. Chapters 4 and 5 are related chapters that comprise profiles of countries active in the climate policymaking arena and evaluate their policy responses in the light of the EM framework presented in Chapter 1. Chapter 3 focuses on countries within the EU, while Chapter 4 investigates five countries that have the common thread of being former British colonies and so have similar legal systems, language and certain common cultural influences. Chapter 5 evaluates the EU as a region while Chapter 6 looks at the UK; these are studied in detail because they provide mature models of climate change management and effective participation in the negotiation of the governing treaties and can therefore provide interesting insights into both climate policymaking and EM theory. Chapter 7 comprises a comparative analysis of the country profiles and case studies. Chapter 8 consists of an evaluation of the perspectives of civil society (employers' organizations, trade unions and other actors) on the governance arrangements for climate change policymaking at the intergovernmental, regional, national, local and other levels. The concluding chapter comments on the analysis and findings of the book and presents some suggestions regarding what actions are needed to improve sectoral involvement in, and uptake of, climate-related policy for policymakers, business and labour.

NOTE

1. Further to the negotiations under the Ad Hoc Working Group on the Durban Platform for Enhanced Action (ADP), the Conference of the Parties (COP), by its decision 1/CP.19, invited all Parties to initiate or intensify domestic preparations for their INDCs towards achieving the objective of the Convention as set out in its Article 2. Without prejudice to the legal nature of the contributions, in the context of adopting a protocol, another legal instrument or an agreed outcome with legal force under the Convention applicable to all Parties. In decision 1/CP.20 it is further specified that in order to facilitate clarity, transparency and understanding, the information to be provided by Parties communicating their intended nationally determined contributions may include, as appropriate, inter alia, quantifiable information on the reference point (including, as appropriate, a base year), timeframes and/or periods for implementation, scope and

coverage, planning processes, assumptions and methodological approaches including those for estimating and accounting for anthropogenic greenhouse gas emissions and, as appropriate, removals, and how the Party considers that its intended nationally determined contribution is fair and ambitious, in light of its national circumstances, and how it contributes towards achieving the objective of the Convention as set out in its Article 2 (UNFCCC, 2015).

1. Ecological modernization: theory and the policy process

INTRODUCTION

Ecological modernization (EM) arose in the early 1980s as a theoretical approach to describing the relationship between economics and innovation, the interventions of the nation-state and the involvement of NSAs in decision-making in order to achieve desired environmental outcomes (Mol and Sonnenfeld, 2000). Some theorists also advocate ecological consciousness as a required element in the EM model, a shift from the implied to the explicit influence over the achievement of outcomes (Howes et al., 2010; Sonnenfeld, 2000; Pellow et al., 2000).

EM is increasingly used in environmental policy analysis (Christoff, 1996; Spaargaren et al., 2009, Howes et al., 2010) because it provides an appropriate framework to explore the roles of actors in society in the process towards achieving best practice environmental outcomes. This book explores whether the public policy process, as framed by EM, should make reference to the employment and workplace impacts of climate change policy and the interventions of employers' organizations and trade unions, notably in climate change-related policymaking.

EM emerges from the social and political sciences and has established its credentials in the environmental policy field, shaping the discourse of environmental politics (Cohen, 1997). Howes et al. (2010) observe that since the 1990s there have been three theoretical approaches that have provided frameworks within the field of environmental politics: administrative rationalism, economic rationalism (neoliberalism) and ecological modernization. They find that ecological modernization as a theoretical avenue has strong relevance to environmental planning and management in many industrialized countries and as such provides a suitable framework within which to explore policy processes.

Much of the discussion concerning EM has been theoretical in nature. In Frederick Buttel's (2000) examination of the ascendance of EM as an influence over environmental policy, he refers to it as a 'well-developed and highly-codified social theory' (p. 58) pertaining to politics and the state. Maurie Cohen (1997) argues that the two major social theories shaping the discourse of environmental politics are Ulrich Beck's (1992) risk society theory and EM. Lange and Garrelts (2007), exploring how risk is introduced into public policy contend that EM is the most effective theoretical approach for solving environmental policy problems. Hampton (2015) and Hayden (2014) discuss EM in terms of its practical application. For Hampton, EM can be used to frame climate change policy quite precisely. Hayden likewise supports EM as an environmental policy approach, which he further links to green growth (Hayden, 2014, p. 5). Both emphasize the role of civil society and labour activists and challenge the relevance of business as usual as a policy option for the contemporary era.

EM theorists acknowledge that research is still required to embed the EM approach in different economic, cultural and political situations (Spaargaren et al., 2009). They also recognize that EM is yet to firmly establish its utility beyond developed economies. Marsden et al. (2011) find that EM has been successfully applied to the development of environmental policy in some regions in China and analysis suggests the results could be broadly replicated in other regions. The work of Sonnenfeld and Mol (2006) in 11 market-oriented industrialized and industrializing states in Asia also finds positive indications of the relevance of EM, although more research is required to confirm its priority over other theoretical frameworks. Partly in response to finding reasons for different EM manifestations in different countries and policy settings, theorists have sought to identify 'weak' and 'strong' types of EM (see for example Christoff, 1996). On this view, 'weak' models of EM might be found in developing states, such as India and Kenya. Although, as discussed below, it is only a partially valid conception of EM, some credence is found for this typology. In the lead-up to the 2015 Paris COP, for example, the governments of both countries submitted qualified INDCs whereby achievement of the targeted outcome of keeping global temperatures within 1.5–2 degrees centigrade of pre-industrial levels was conditional on the provision of significant financial support and technology from sources external to the country, since the institutions of the state are fragile and priorities are oriented more towards economic development and poverty eradication than emissions reduction per se. Both countries are vulnerable to the effects of climate change and, while EM helps identify the possible shortcomings of the proposed policy, to a large extent it is beyond the capacities of the governments and NSAs to pursue a path of 'strong' EM. Conceptions of EM such as these represent one of several ways of classifying EM, which has gone through several stages of theoretical development. The evolution of EM is shown in Figure 1.1.

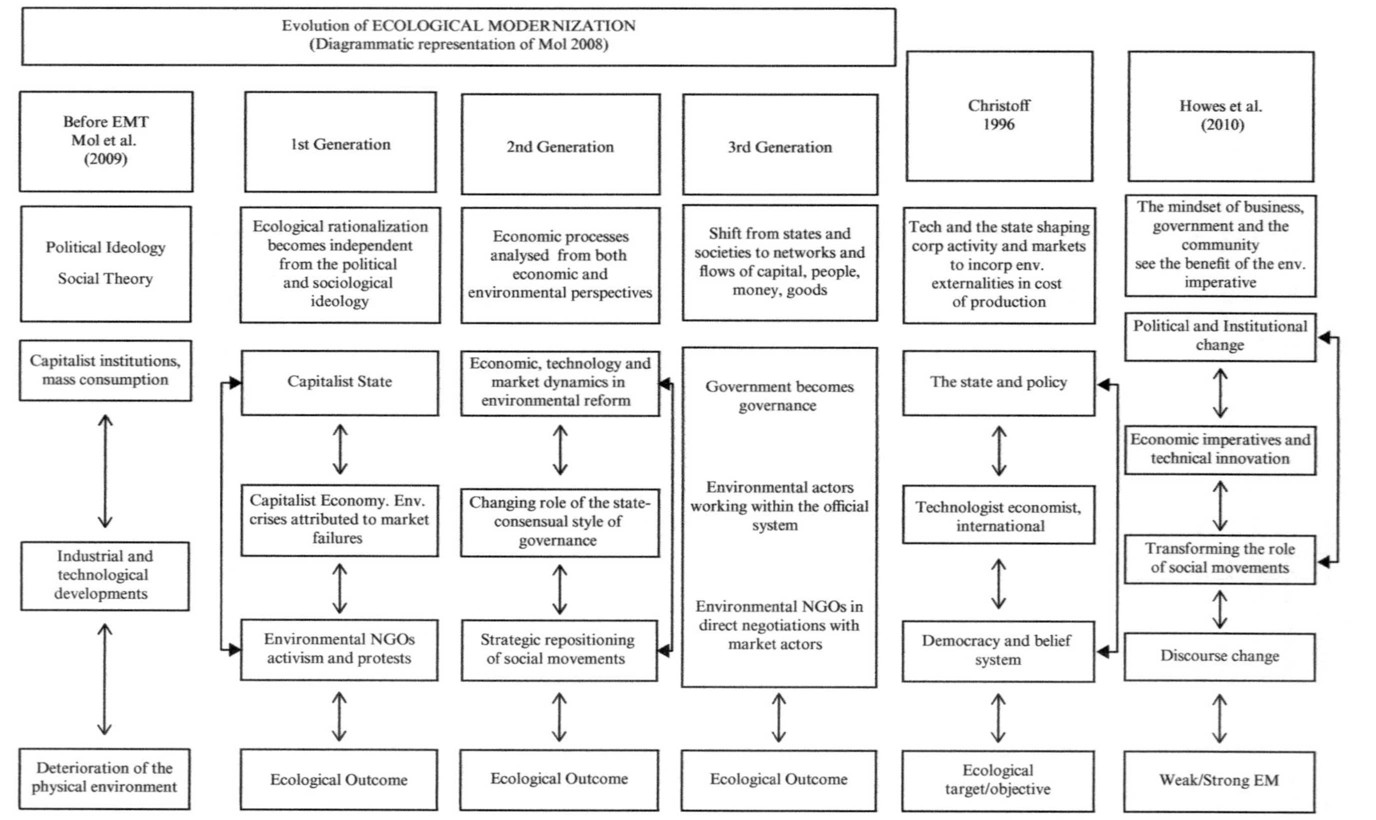

Figure 1.1 Evolution of ecological modernization

Investment, Innovation and the Interventions of the Nation-state

EM contends that the relationship between economic activity and innovation in technology and the interventions of the nation-state and civil society are required to achieve best practice environmental outcomes. Joseph Huber (1991) emphasizes the role of the state in EM, advocating that the legal foundations of environmental policy and regulation by authorities are absolutely indispensable and provide the stability factor necessary for business decision-making processes to innovate. He contends that the nation-state is the essential counterbalance to the unfettered behaviour of the market and its role as an active regulator is fundamental to effective environmental policy. Huber's (1991) views are supported by, among others, Hajer (1995), Sonnenfeld (2000) and Cohen (1997). However, the nation-state and industry are not mutually exclusive and it is argued by Ashford (2002) that the optimum situation is one where the nation-state provides clear standards and policy goals while allowing flexible means for industry to achieve its goals.

While articulated in different ways by theorists, there is a consistent thread that binds the contribution of enterprise with strict monitoring of performance by the state. Stern (2007) contends that effective climate change adaptation may require governments to address specific market failures and barriers. Esty and Porter (2005) concur, observing that the market on its own would not implement the strategies necessary to deliver environmental outcomes without a regulatory push. They find that environmental results are not merely a function of economic development but also a consequence of policy choices. They conclude from their research that, amongst other things, regulatory stringency and regulatory structure are highly significant in the achievement of environmental outcomes. These observations deviate significantly from Porter's earlier (1990) views, in which he contends that experimentation by the nation-state with policies to promote national competitiveness 'usually ends up only undermining it' (p. 73) and, by inference, that industry left to its own devices will deliver a competitive economy. Porter later observed that the market 'makes a false assumption about competitive reality – namely, that all profitable opportunities for innovation have already been discovered, that all managers have perfect information about them, and that organizational incentives are aligned with innovating' (quoted in Porter and van der Linde, 1995, p. 127). Stern (2007) makes a similar observation, that 'in the absence of public policy there are limited or no returns to private investors for [serving the public good]' (p. 27). Such contentions indicate a belief that the market on its own will not make the decisions to invest and innovate to develop the technologies necessary to meet the challenge of reducing GHG emissions. They also acknowledge that regulation on its own does not inspire investment and innovation. The progress towards specified ecological outcomes relies on a cooperative and mutually beneficial relationship between the nation-state with actors in the economy and investors in innovation.

Civil Society

The activists in the EM process are variously described as non-governmental organizations (NGOs) and civil society. This difference in terminology can be best understood by drawing a distinction between actors and their respective roles, a distinction that is discussed more fully in the following chapter. To recap, civil society is the collective term for civil society organizations (CSOs) and is defined as 'the wide array of non-governmental and not-for-profit organizations that have a presence in public life, expressing the interests and values of their members or others based on ethical, cultural, political, scientific, religious or philanthropic considerations' (World Bank, 2013, p. 1). 'Non-state actor' is a term used to refer to non-governmental organizations, which may be civil society organizations, transnational corporations and intergovernmental organizations (Weiss et al., 2013; Joey, 2015).

When EM theorists refer to civil society, they are alluding to environmental activists. In his discussion of EM theory, Mol (2008) uses the terms NSAs and NGOs as descriptors of activists and institutionalized players such as Greenpeace, Friends of the Earth, the World Wide Fund for Nature and the 'anti-globalisation movement' (p. 199). Although Mol does not use the term 'civil society', the groups he identifies are united by their environmental activism. Gonzales (2009) similarly infers that civil society means activists, explaining that civil society is a 'more democratic venue than the state because it is relatively unconstrained' (p. 211) and that as civil society the actors are not subordinated to state policy. Dryzek et al. (2009) offer a more expansive and inclusive role, discussing the interaction among a broad array of political, economic and social institutions resisting subordination of ecology to economics.

While scholars may generally speak of civil society as consisting largely of environmentalists, activism on environmental and climate issues in civil society has a wider scope in practice. UNFCCC has given formal recognition to more than 1,300 civil society organizations, including business and industry, environmental groups, farming and agriculture, indigenous populations, local governments and municipal authorities, research and academic institutes, labour unions and women, gender and youth groups (UNFCCC, 2011a). In its own wisdom UNFCCC refers to these all as NGOs and a whole series of related acronyms have arisen, of note to this study being BINGOs (business and industry), TUNGOs (trade unions), as well as environmental NGOs (ENGOs) (UNFCCC, 2011a). Golmohammadi (2012) conceives of civil society as a collective, referring to the United Nations Major Groups as their representative organizations. He proposes a management model for their collective representation as similar to the ILO, in which governments and the elected business and worker representatives have equal rights and equal representation. In the EU system, civil society is understood in the broader context and is afforded a statutory role in the formal policy process. Article 300 of the Treaty that forms the EU provides that 'the European Parliament, the Council and the Commission shall be assisted by an economic and social committee (EESC) exercising an advisory function and . . . shall consist of representatives of organisations of employers, of the employed and of other parties' representative of civil society' (European Union, 2010, Art. 300). The Treaty states that the institutions of the EU must consult with the EESC on matters that include the environment and climate change (Europa, 2012b).

Earlier in the emergence of EM as an accepted theoretical framework in environmental policy analysis, it was said that 'little is known still on how, to what extent and how successfully environmental interests are included in all kind of economic, cultural and political practices' (Spaargaren et al., 2009, p. 511). More recently, international agreements (2012 Rio + 20 Summit; the subsequent COPs and most recently the 2015 UNFCCC climate treaty known as the Paris Agreement) have embraced the potential of civil society, aware of the need to harness every opportunity to reduce GHG emissions. That said, national and international policy processes remain confused in the conceptions of civil society, which can have a negative impact on its ability to participate in policymaking. On the eve of the 2015 Paris COP and in the wake of the fatal terrorist attacks that occurred on 13 November that year, civic authorities placed a ban on civil society marches and protest actions. Although this ban was later lifted it had a chilling effect on non-state participation in the climate talks. This action also confused a range of civil society protagonists one with another (anti-capitalist activists with advocates for increased climate action), and also conflated proponents of non-violent direct action with actors explicitly outside civil society, such as anarchists (Cadman, 2015).

The International Influence

A series of events focused on global approaches to the sustainable development, such as the Brundtland Commission of 1987 (WCED, 1987) and the Rio Summit of 1992 (UNCED, 1992) as well as the wave of environmental events that followed created the expectation of collective international action concerning climate change. However, although the scope of these events was global, they are outcomes to which the Parties (the governments) have committed and accordingly also had significant influence over domestic environmental policy. In this regard, Spaargaren et al. (2009) warned of the potential conflict between international and domestic institutions, observing that the division of tasks may cut across the institutions of the state, market and civil society by mandating the sharing – and possible delegation – of responsibilities to international institutions.

Few would argue the validity of the concerns expressed by Spaargaren et al. (2009), yet in practice there are signs the model established to deliver the commitments under international agreements is failing and that domestic stakeholders feel disenfranchised. The UNFCCC management of the 2009 COP 15 in Copenhagen was criticized because it could not facilitate civil society participation in the dialogue and diminished the opportunity to civil society for access to officials (Fisher, 2010). This, with the inability of the subsequent COPs to agree to increased action to mitigate climate change or commit to the next generation of international climate agreements, has seen the focus of civil society action shifting away from such international events. Eastwood (2011, p. 35) observes that 'government participants and civil society actors alike are starting to admit that, as international negotiations stall out on the global level, local actions and smaller-scale projects will be the locus of climate change mitigation and adaptation'. While there were criticisms due to the lack of progress (BusinessEurope, 2012: Ryan, 2012), civil society maintained an active presence in the process.

By 2015 the Kyoto Protocol had effectively reached its use by date (beyond certain provisions under the Doha Amendment) and Paris loomed as the final deadline for an agreement, with further extensions unlikely. The Parties were finally able, after two weeks of tense negotiations, to find the common ground necessary to deliver a framework suitable for the move to the post-Kyoto environment, based around provisions of the Lima Paris Action Agenda (UNFCCC LPAA, 2016). The Paris Agreement permits a greater degree of flexibility for nation-states in policy and programme development and outcome delivery than under the Kyoto Protocol factors that can better facilitate the necessary domestic and international coexistence and deliver the targeted ecological outcome of reducing GHG emissions and containing the rate of global warming to two degrees or less.

OPERATIONALIZING ECOLOGICAL MODERNIZATION

The importance of EM theory to climate change policy has been well established; however, as outlined previously there are still limits to its applicability. Sonnenfeld (2000) argues that before the applicability of EM is examined, it must first be formally defined and operationalized. While operationalizing can be deemed as simply putting EM into practice, the process of defining EM is more complicated.

There is evidence of the increasing influence of EM in policy development. Hayden (2014) observes that in 2008 61 per cent of voters in Canada backed parties proposing some version of an EM-type climate response and in 2011, voters favoured pro EM policies. Hampton (2015) adds that the UK Environment Secretary in 2006 appealed for trade union engagement in climate change in ecological modernist terms. However, EM is still far from operational in the sense that the process to operationalize is not standardized and agreed upon. To reiterate the Christoff (1996) analysis of EM, the competing definitions in EM are a factor that inhibit its operation and there are aspects of EM that are interpreted differently within the literature. This flaw in the EM process has left theorists open to the criticism that they have not shown that using EM as a methodological framework leads to ecological improvements or transformation or that it reduces the direct impacts on the environment. York and Rosa (2003) contend that theorists place a greater focus on the institutional transformations than the consequences of those transformations, a contention that is strengthened if the process of operationalizing the theory fails.

Measuring Policy Effectiveness

A review of the effectiveness of climate change policy measured within the framework of EM may provide a reference point from which a methodology to operationalize the framework can be developed and qualitative and quantitative research approaches can be applied. Esty and Porter, for example, identified a benefit to environmental policy analysis and developed an environmental performance index (EPI) using both qualitative and quantitative measures (Esty and Porter, 2005). This and other scales have been used to determine 'weak' and 'strong' environmental performance. Tables 1.1, 1.2, 1.3 and 1.4 present a variety of these. Collectively, 'strong' EM is portrayed as a collaborative, deliberative process, where the state provides markets and civil society actors access to all the necessary information and institutions for achieving the desired environmental outcomes. The state then regulates and acts to ensure compliance. Moderating themes across the different conceptualizations are that some 'strong' versions allow for multiple EM possibilities, while others focus more on EM's contribution to more internationally oriented outcomes. 'Weak' ecological modernization is represented as less collaborative, the state regulates to facilitate the market, the focus is national rather than international and the strategy is hegemonic rather than offering multiple opportunities.

Table 1.1 Ecological modernization as viewed by Howes et al. (2010)

	'Weak' EM	'Strong' EM
View of the environment	Economist and utilitarian	Ecological
Role of the state	Market facilitation, information dissemination, minimum state intervention	Substantial state intervention, institutional restructuring, reforms to economic and regulatory policies
Policy approach	Instrumental	Communicative
Decision-making style	Technocratic/closed decision-making by economic and political elites	Deliberative democratic/open, with participation and involvement
Scale of focus	National focus on developed nations	International
EM strategy	Hegemonic	Diversifying, multiple possibilities with EM providing orientation

Source: Howes et al. (2010).

Table 1.2 Ecological modernization as viewed by Christoff (1996)

'Weak' EM	'Strong' EM
Economistic	Ecological
Technological	Institutional/systemic
Instrumental	Communicative
Technocratic/neo-corporatist	Deliberative democratic/open
National	International
Unitary	Diversifying

Source: Christoff (1996).

Janicke's (2008) EM-friendly framework of environmental regulation (shown in Table 1.4) is a fair representation of the vast middle ground between these representations of 'weak' and 'strong' conceptions but has a clear leaning to 'strong' EM. Mol (2001), in his comparison of EM with the treadmill of production concept, also sits in the middle but with a leaning to the 'weaker'.

Table 1.3 Environmental change versus economic continuity

	Treadmill of production	Ecological modernization
Kind of radicality	Economic radicality	Environmental radicality
Environmental improvements	Absolute sustainability	Relative improvements
Assessment of environmental change	Window dressing	Real changes
Relationship between changes analysed and changes proposed	Weak relation	Strong relation
Main emphasis	Institutional continuity	Institutional transformations

Source: Mol (2001).

Table 1.4 Ecological modernization and innovation-friendly framework of environmental regulation

Policy objective	Policy content
Instruments are EM and innovation friendly if they	Provide economic incentives
	Act in combination
	Are based on strategic planning and goal formulation
	Support innovation as a process and take account of the different phases of innovation/diffusion
A policy style is EM and innovation friendly if it is	Based on dialogue and consensus
	Calculable, reliable, and has continuity
	Decisive, proactive, and demanding
	Open and flexible
	Management oriented
A configuration of actors is EM and innovation friendly if	It favours horizontal and vertical policy integration
	The various objectives of regulation are networked
	The network between regulator and regulated is a tight one
	The relevant stakeholders are included in the network

Source: Janicke (2008).

While these measurement tools are appropriate as descriptive representations of policy rating options within the framework of the theory, they do not sufficiently inform the policy development process and the interventions necessary to provide certainty about the environmental outcome. The terms 'weak' and 'strong' are taken from the literature and are used here in the absence of other understood nomenclature, or a quantitative measure such as that provided in the EPI. Mol (2001) holds the view that EM is about relative improvement in environmental conditions rather than absolute sustainability. Christoff (1996) argues that 'weak' and 'strong' are measured by the choices made. Hajer (1995) speaks of EM as a discourse that recognizes the structural character of the environmental problematic and institutions can internalize the care for the environment. Howes et al. (2010) speak of a situation where ecological principles are seen by the state, market and civil society actors as the required policy goal of both institutions and business.

A discussion about measures of effectiveness should focus less on whether EM is 'weak' or 'strong', implying 'good' and 'bad' but, rather, if it is able to account for the outcomes that are expected from the policy initiatives. A policy is structured to deliver a specified result, and may yield an 'ideal' outcome, free from any constraining factors – or the outcome may be contingent, such as one determined by a nation-state's unique set of circumstances. In reality, the ideal is rarely achievable. The outcome may therefore arise from the best available (or 'optimum') set of policy choices, given contingent factors. Furthermore, given such contingencies, it is also possible that a given set of policy choices may not represent the best available set of choices, and be influenced by further factors. In this case, the outcome may be 'sub-optimum'.

For the purposes of the discussion here the outcome sought by the state is a consequence of its sustainability and climate change policies. In the EU these are based on its 2030 Climate and Energy Framework, while in the UK it is its Climate Change Act 2008 and related legislation. In the case studies that follow, the findings are analysed to determine what version of EM model they conform to, what the policy framework constitutes, and whether it represents an ideal, optimum, or sub-optimum outcome.

Climate change policy is influenced by a range of contingencies, which includes national, cultural, economic, social and political contexts. Within these parameters there are no absolute measures, that is, actions that are correct in all situations, merely what is achievable and appropriate to the circumstances. Consequently, it is suggested that while useful parameters for general assessment, the 'weak' and 'strong' terminology is not sufficiently nuanced to represent the gamut of choices, which policymakers confront, and represents a dualist classification, that does not fully align with contemporary social-political circumstances. Therefore, it is used here for descriptive purposes of the existing EM literature, rather than constituting a preferred typology.

Operationalizing Ecological Modernization

To implement EM and permit EM to guide policymakers requires consistency in definitions and terminology. Christoff (1996) argues that the competing definitions in EM is a factor that inhibit its operation and there are aspects of EM that are interpreted differently within the literature, some of which are noted in Table 1.2. Additionally, the application of EM is hampered by a lack of clarity about the preferred steps to guide the process of its operationalization.

The multiple descriptions of EM are a rich source of diverse opinion about its true nature. A list of some of the theorists' conceptualizations of EM is shown in Table 1.2. These conceptualizations consider EM as either a general descriptor of trends (Cohen, 2000), a discourse that recognizes the structural character of the problematic (Hajer, 1995) and a centripetal movement of ecological interests, ideas and consideration in the State's institutional design (Mol, 2008). Ashford (2002) adds the requirement for the involvement of a broad array of stakeholders. This research concludes that the varied conceptualizations of EM do not blur its usefulness, a concern expressed by Christoff (1996), but rather add to the complexity of the task of its operationalization.

Across all of the differing views of what EM is and does, each of the conceptualizations are built around the elements of innovation, the state, market, civil society and ecological consciousness. As shown in Table 1.5, the language of EM theorists provides potentially conflicting interpretations of EM and its potential to be operationalized. This leads to the understanding that if EM is to serve as a guide in the practice of policy development, it is necessary to find consistency in language and understanding. The standardized description of EM proposed for this research is the relationship between the innovation and technology, the market, interventions of the state, the role of civil society and ecological consciousness in decision-making in the pursuit of environmental outcomes.

The theorists' discussions and social theory literature do not apply the language of EM consistently and often use different terms when describing the same concepts. This creates confusion and establishes a potential barrier to the theory's operationalization. For instance, the term 'institution' is sometimes used as a reference to institutions of the state, while in other discussions it is a reference to institutions of the market. In order to address the problem created by the multiple interpretations, a

Table 1.5 Conceptualizations of ecological modernization

Theorist	Definition of EM
Cohen (2000)	Defines EM as a term adopted as a general descriptor of trends in the application of science and technology to environmental problems and the efforts to reconcile conflicting objectives between responsibility for the environment and continual economic expansion
Hajer (1995)	Conceptualizes EM as the discourse that recognizes the structural character of the environmental problematic but nonetheless assumes that existing institutions can internalize the care for the environment
Howes et al. (2010)	Speaks of a situation where ecological principles are seen by the state, market and civil society actors as the required policy goal of both institutions and business
Janicke (2008)	Considers EM a systematic eco-innovation and its diffusion
Langhelle (2000)	Conceives of EM not so much a set of sociological theories but rather as a political programme favouring a particular set of policies
Mol (2001)	Contends that the notion of EM can be seen as the social scientific interpretation of environmental reform processes at multiple scales in the contemporary world. EM studies reflect how various institutions and social actors attempt to integrate environmental concerns into their everyday functioning
Mol (2008)	States that the basic idea of EM is that at the end of the second millennium, modern societies witness a centripetal movement of ecological interests, ideas and considerations in their institutional design
Mol and Sonnenfeld (2000)	Refer to EM as the relationship between economics and innovation, the interventions of the nation-state and the involvement of NSAs in the decision-making to achieve environmental outcomes
Simonis (1989)	Submits EM as resolution of environmental problems through harmonizing ecology and economy
Ashford (2002)	Tenets of the still-evolving EM theory that have been melded into the following theory: Unregulated capitalism is responsible for the present ecological and environmental problems and this is partly because the prices of goods and services do not adequately represent the social cost of production and consumption

Table 1.5 (continued)

Theorist	Definition of EM
	Under thoughtful reflexivity, the present and enlightened industrial actors can succeed in advancing the material wellbeing of citizens, contribute to their nation's competitiveness and can also contribute to the necessary scientific and technical changes (innovations) in products, processes and services to adequately meet the environmental challenges, especially if a broad array of stakeholders is involved

standardized definition is proposed for each of the terms, while remaining true to the intent of the theorists. These definitions are discussed below.

Proposed definitions

- *Innovation and technology* is adopted as the standardized term to represent the intentions of the theorists when they speak of science (Cohen, 2000), science and technology (Mol and Sonnenfeld, 2000) and technological innovation (Howes et al., 2010; Ashford, 2002). The meaning adopted for this research defines innovation and technology as 'the application of new solutions and the practical application of knowledge' (Merriam-Webster, 2013).
- *The state* is also variously described in EM, with some authors focusing on particular attributes such as the state as the regulator, change agent and institution (Howes et al., 2010; Janicke, 2008; Dryzek et al., 2009). This research defines the state as the agent for shaping the structures and policies of the state and other national and local actors in social life including business, politics and civil society (Meyer et al., 1997).
- *The market* is variously described as pursuit of the economic imperative (Howes et al., 2010) and application of economic rationality (Dryzek et al., 2009). The meaning adopted for this research is the market as the area of economic activity in which buyers and sellers come together and the forces of supply and demand affect prices (Merriam Webster, 2013).
- *Civil society* is described in some instances quite differently (see section Defining Civil Society in Chapter 2 for a more detailed discussion). The discussion seeks to reconcile the EM theorists'

general representations of civil society as environmental activists with the Organisation for Economic Co-operation and Development (OECD), UN and World Bank's broader considerations of civil society as a range of stakeholders and interest groups that are involved and influential over public policy. The meaning adopted for this research is civil society as the wide of array of organizations including community groups, NGOs, labour unions, indigenous groups, charitable organizations, faith-based organizations and professional associations (World Bank, 2013).

- *Ecological consciousness* was selected over other terms that are used in EM such as discursive change (Mol and Sonnenfeld, 2000; Howes et al., 2010), Christoff's (1996) hegemonic progress and Cohen's (2000) epistemological perspective. The meaning adopted for this research of ecological consciousness is the consideration of ecological principles in the desired policy goals of both institutions and business (Howes et al., 2010; Christoff, 1996; Cohen, 2000).
- *Ecological outcome* is discussed at length in the following section and introduces the concept of defined pathways to provide a more concrete guide than the range of weak to strong EM that is the language of many theorists.

The standardization of these terms removes one of the barriers to the operationalization of EM. Consequently, the process to develop policies that implement EM is provided here. This represents an ideal EM. However, such an EM type is not what is always sought by policymakers, who may prefer a template that simply allows for the monitoring and evaluation of contingent policy choices. An ideal EM framework, it is suggested, constitutes the following:

1. Orientation to EM theory. Policy objectives seek to harmonize the economy with the environment; standardized definitions to strengthen governance of EM and its components exist; application of the policy is ensured by sufficient resources and related capabilities; policy application and enforcement includes the achievement of the specified ecological outcomes;
2. Outcome selection. The desired outcome of the policy is specified – these may be science based around addressing specific environmental concerns, a process-oriented model that seeks to harmonizes the economy and the environment in specific but realistic ways, or the achievement of a specified outcome such as a GHG emission reduction target;
3. Policy interventions that deliver the desired outcome.

ECOLOGICAL MODERNIZATION THEORY: THE ECOLOGICAL OUTCOME

The Pathways to the Environmental Outcome

Outstanding from the discussion to standardize the definitions in EM is a definition of what constitutes the 'ecological outcome'. The specification of an ecological outcome will take into account whether the measure of harmonization of the economy and the environment is to be judged in the absolute or whether it is a measure of what is achievable in the particular circumstances of a country, as alluded to previously. In many countries, the outcome specification reflects the commitment by the state to international climate change agreements. In Europe the outcome specification is the EU 2030 Climate and Energy Framework (Europa, 2016) to deliver reduced emissions and energy security, which embraces the EU commitment to UNFCCC agreements.

Analysis of theorists' conceptualizations of the outcomes in the EM model led to the conclusion there is not one clear outcome but a range from the environmental imperative to that of decoupling economic growth from environmental harm. This range, commonly described as 'weak' to 'strong' EM outcomes, can be described more precisely across the range as:

- Outcome 1: harnessing science and the environment to address contemporary ecological concerns;
- Outcome 2: decoupling economic growth from environmental harm where success is measured as progress towards absolute sustainability;
- Outcome 3: decoupling economic growth from environmental harm where success is measured as progress towards a specified outcome or a target.

Table 1.6 shows the ecological outcomes aligned to these three pathways. Table 1.7 then charts the possible interventions along the pathways to deliver the outcome in the context of the components of EM.

The pathway to Outcome 1 reflects an environmental requirement to which other considerations are subordinate. The policy framework to support these objectives could involve innovation from science and technology, institutional reshaping by the state and transforming ecological consciousness through the reflexive process of social learning. Policy interventions could be waste reduction and elimination, resource recovery and reuse, with long-term objectives as resource conservation and clean production[1] (Sonnenfeld, 2000; Janicke et al., 1997; Hajer, 1995; Cohen, 2000).

The pathways to Outcomes 2 and 3 rely on a common policy framework described by Maarten Hajer (1995) as making environmental degradation calculable, environmental protection as a positive sum game and the reconciliation of economic growth with ecological systems. The policy interventions proposed by Howes et al. (2010) include technological innovation, engaging with economic imperatives, political and institutional change, transforming the role of social movements and ecological consciousness. Although the policy framework for Outcomes 2 and 3 are common, the pathways differ in their focus. The pathway for Outcome 2 has a focus on process-oriented policy, whereas the pathway for Outcome 3 addresses the policy choices to deliver the outputs required.

A range of approaches to environmental policymaking and for assessment of sustainability have been provided in research and reports by the OECD (2013a), Green Growth Knowledge Platform (GGKP) (2013) and the World Bank (2013). The GGKP has also developed a comprehensive set of diagnostic indicators that have parallel features with the EM model.

The GGKP is a global network of researchers and development experts sponsored by the Global Green Growth Institute, OECD, the United Nations Environment Programme (UNEP) and the World Bank that identify and address knowledge gaps in green growth theory and practice. The GGKP research is a model of Outcome 2 and speaks of the pathway in the five well-recognized stages of the public policy development process: agenda setting, policy formulation, decision-making, policy implementation, and monitoring and evaluation (GGKP, 2013). Policy formulation is also best guided by a process that is evidence based, which means that it has rigour, is inclusive and can be evaluated. The evidence-based approach of EM is discussed in the next section.

In EM terms, the OECD proposal is a pathway to Outcome 3. The OECD (2013a) proposes policy instruments in three categories: taxing, pricing and mechanisms that value natural assets; regulations, standards and information policies; and a set of cross-cutting policies to stimulate green growth in a systemic way. The cross-cutting policies referred to are investment, research and development; labour and skills; and climate adaptation (OECD, 2013a; 2013b). It reports that progress along the pathway to achieve a chosen outcome requires commitment to a vision and plan for green growth. The process to achieve the outcome involves the design, reform and implementation of policies that stimulate green growth, and strengthen governance – including capacity building for sound decision-making to monitor and enforce policies.

Two points in the OECD (2013a) approach merit comment here. First, labour and skills are important issues to be included in both short and long-term climate change planning; second, targets are beneficial in climate change planning but if there are too many there is the potential for conflicting policy, a consideration the UK has not factored into its climate change strategies. The UK has set itself many targets, making commitments to international and EU agreements and its own further targets. An OECD (2010) review of UK climate change policies added that more could be done to align the economic and environmental objectives and the integration of programmes across portfolios. Additionally, monitoring of climate policy effectiveness by the UK government's Committee on Climate Change has found that some programmes are at the limits of their capacity and that step change is required in the suite of policies if all of the set targets are to be met (DECC, 2011e). The OECD and ILO reviews of the UK labour market and skill development finds the policies are weak and the anticipated shortage of labour and skills will create barriers to service delivery (Miranda and Larcombe, 2012; Strietska-Ilina et al., 2011; Gleeson et al., 2011).

The finding that there is not one single scenario to be considered in the operationalization of EM but a range is critical, as it is only by committing to an outcome that the state can make the effective policy interventions that will lead to the required environmental outcome. The choice of outcome will dictate whether policy requires a strong science orientation, whether policies will be seeking to deliver progress towards absolute sustainability or whether they deliver a target such as specified volume of GHG emissions reduction.

Table 1.6 Theorists' objective for ecological modernization

	Outcome 1	Outcome 2	Outcome 3
Huber (1991)		X	
Hajer (1995)	x		
Christoff (1996)		X	
Cohen (1997)	x	(x)	
Jokinen (2000)			x
Pellow et al. (2000)			x
Sonnenfeld (2000)	(x)		x
Mol (2001)			x
Janicke (2008)		X	
Howes et al. (2010)		x	

Note: (X) Primary influence over the outcome; (x) Secondary to the primary influence over the outcome.

Table 1.7 Scenarios for the achievement of ecological modernization

Pathways		Outcome 1	Outcome 2	Outcome 3
Innovation and technology	a			x
	b	x	x	
State	a			x
	b	x	x	
Market	a		x	
	b			x
Civil society	a			x
	b* Other			
	Environ		x	
Ecological consciousness	a			x
	b	x	x	

Note: a and b: Indicates the relative priority where b is the higher priority; *Other and Environ draw a distinction between other civil society organizations and environmental activists.

Evidence-based Policy

This section presents an approach to public policy development within an EM framework that is based on evidence and processes that have rigour. The GGKP (2013) contends that green growth and green economy policies need solid evidence-based foundations. EM introduces policy considerations that are not necessarily informed by hard physical data but are qualitative by nature, such as the role of civil society, or the value to policy implementation that is supported by the community.

Evidence-based policy (EBP) is public policy informed by rigorously established objective evidence (Head, 2010). While it is not the only process for developing policy, it is a process that demonstrates evidence from qualitative study can be accepted as objective provided that it is collected in a manner that has discipline and where the outcomes can be measured. While an EBP approach is traditionally based on quantitative methods and statistical techniques and analyses, the value of qualitative evidence has been widely debated in the EBP community in recent years (Smith, 1996). The World Health Organization (WHO) has developed guidelines using evidence-based approaches that rely on qualitative data gathered from existing research, input from multidisciplinary experts, and peer-review (WHO, 2011). In the UK, the Cabinet Office maintains that evidence for the purposes of informing policy includes the management of academic research and professional and/or institutional experience (Parsons, 2002).

There are three enabling factors that underpin modern conceptualizations of EBP: high-quality information based on relevant topic areas, cohorts of professionals with skills in data analysis and policy evaluation and political incentives for utilizing evidence-based analysis and advice in decision-making processes. Western countries have developed a strong institutional foundation for nurturing EBP capacities. Their commitment to good data and sound analysis has been reinforced by their involvement in international organizations such as the OECD and their endorsement of international agreements that require sophisticated reporting.

However, while the institutional capacity necessary to support EBP has increased, public policy development in the Western countries is still often based on guesswork and assumptions rather than as a process in which the social policy or sciences have an influential part to play (Parsons, 2002; Hayes, 2007). In the UK in 1997, the newly elected Labour government sought to address that situation and in its White Paper *Modernizing Government* declared its intention to introduce evidence-informed processes in order to provide better-informed development and delivery of government policy (UK Government, 1999). To ensure the requirements of the White Paper would not stall for lack of appropriate capabilities within the public service, the government provided for access to skills, experience and domain knowledge developed outside the government that could be used to improve the quality of policy discourse within government departments (Levitt and Solesbury, 2005). Interestingly, it was not monitored or evaluated following the Paper's release, and leads to questions of whether in fact it is being effectively implemented, a reasonable question when the government's Committee on Climate Change reports evidence of resistance to adoption of climate change policies within the public sector (DECC, 2011f).

In EM, theorists have traditionally been more concerned with developing the conceptual and policy implications of their framework and less with evidence-based analysis. If EM is to be relevant and used as a guide in the development of contemporary policy, the pathways to the chosen outcome must provide for an evidence-based policy selection.

Monitoring and Evaluation

The monitoring and evaluation phase of policy development and implementation is essential for assessing the need for policies and whether they achieve their stated goals. The evaluative criteria developed by Guyatt et al. (2008) rates the quality of the evidence on a scale of between high, moderate, low and very low quality. WHO's (2011) policy indicator and guideline development process commences with the formation of a Guidelines Review Committee to undertake a scoping (including a review of existing guidelines and selection of the critical outcomes and draft key questions); the formation of, and consultation with, an expert panel; implementation of a step-by-step methodology; meeting of the global panel of experts; final guidelines report; clearance by the Guidelines Review Committee; and finally, publication and dissemination.

The Kyoto Protocol introduced a platform for a monitoring, evaluation and reporting regime, which was a significant achievement and an important requirement in effective public policy (UNFCCC, 1997). The Paris Agreement as a 'next generation' treaty builds on this base to implement a comprehensive programme that is aimed at ensuring transparency of action and support, as well as facilitating the reporting of progress towards achieving targets in a timely manner – and undertaking appropriate remedial action if they are not (UNFCCC, 2015, paragraphs 85–105). To that end, an EM model must also accommodate these requirements.

ECOLOGICAL MODERNIZATION AS A GUIDE TO POLICY CHOICES

Can an operationalized EM guide the policy development process and the policy choices? The negotiation of international agreements is a complex and lengthy process that demonstrates the commitment of the states to the environment and in particular to addressing climate change, a commitment that it is expected will guide national approaches to domestic policy development. Agreements are not currently framed by theoretical models and an operationalized EM might well serve international and domestic policy development processes as they adapt to the economic and social changes necessary to reduce GHG emissions. But it also needs to be recognized that each nation-state has unique political, economic, social and cultural sensitivities that influence policy choices, and these may be constraining factors over implementation of 'stronger' versions of EM; as required to achieve the optimum EM outcome for the given context.

The agreements from the Rio + 20 Conference (UNCSD, 2012a) and the UNFCCC COPs influence the decisions of the state and the policy choices for climate change adaptation and mitigation strategies. To establish if there were indications that there may be gaps between the EM framework, UNFCCC agreements and domestic climate change policy, the EM template presented in this volume was overlaid on a sample of countries notable for their contribution to global climate change management efforts as well as being high GHG emitting countries.

Table 1.8 is a diagrammatic representation of the findings from this test.

While EM theorists generally speak of EM on a scale of weak to strong, it was considered that policymakers would be better assisted by a more definitive measure. To that end and with consideration of the Pathways to operationalize EM, measures of low, medium and high were adopted in place of the weak to strong scale, and the values of 0, 5, and 10 were applied to low, medium and high (weak to strong) models of EM. In some instances, the policy options were conditional, for example in the UK the commitments by the state are ambitious and are supported by a suite of programmes intended to deliver the targeted outcomes and are accordingly are afforded a high (10) rating. However, the government's Committee on Climate Change had submitted to the government that many of the programmes are at their capacity and must be scaled up if targets are to be met. The government to date not acted on this recommendation, a situation that it is considered must be reflected in the valuation as a negative rating (−10), providing range in the model to guide policymakers.

The spread in the scores is indicative of the diversity of the situations, which are in each case framed by policy designed to achieve the expressed outcomes. For the UNFCCC, the outcome of containing global warming to less than 2 degrees while also facilitating social development and economic growth significantly influence the policy options and is the reason for a high spread in the scores. The EU and UK introduced an energy security mandate, and China and India are still pursuing a development agenda. In the scoring of the respective agreements and policies, the EU performance was strengthened by the effective implementation of policy by its member states, as demonstrated in the case of the UK. UNFCCC performed relatively well in terms of achieving its desired outcome. All three demonstrate a solid performance. China and India did not achieve their optimum EM outcome. Of note in all these results is the role extended (or withheld) from civil society actors. In EM, civil society is integral to the achievement of the ecological outcome.

Overlaying the EM template on the chosen sample identified where the agreements and policies diverge from the EM model. For example, the UNFCCC seeks to achieve the multiple objectives of containing global warming, promoting economic growth and social development; the EU and UK have added energy security as an imperative, and in India and China the institutions of the state, market and civil society are weak. Until the 2015 Paris COP, the UNFCCC agreements also exempted some of the world's major GHG emitters (China, India, Brazil and Indonesia) from legally binding commitments to reduce

Table 1.8 Ecological modernization template

	Ecological modernization	UNFCCC	European Union	United Kingdom	China	India
Targets/policy objectives	Decoupling economic development and environmental protection	Global warming ≤2°C Economic growth Social development	40% GHG reduction by 2030	EU package, plus reduce GHG emission 80% by 2050	Reduce GHG emission by 45% 2050	Reduce emissions intensity of GDP 25% by 2020
Market	10	10/−10[4]	10	10	10/−5[5]	5/−5[6]
Innovation and technology	10	10	10	10	10	5[1]
State	10	10	10	10/−5[2]	5/−10[3]	5/−10[3]
Civil society	10	10/−5[7]	10/−5[8]	10	0[9]	0[9]
Ecological consciousness	10	10	10	10/−5[10]	5[11]	5[11]
Ecological outcome	50	50/−15 = 35	50/−5 = 45	50/−10 = 40	30/−15 = 15	20/−15 = 5

Notes:
1 Government and markets focus is on other priorities
2 Government's Committee on Climate Change finds programmes insufficient to meet long-term targets
3 Qualified commitment by the state/prioritizes economic development and poverty alleviation ahead of climate change
4 Market initiatives effectiveness diminished by conditions imposed
5 Strong support for the green market but poor regulations, measurement and controls over green compliance
6 Market is only partially committed to green opportunities, weak institutions of the state, poor regulations, measurement and controls over green compliance
7 Still to engage effectively with civil society
8 Statutory role of EESC introduces stakeholders without necessarily demonstrating a stake in the outcome
9 No effective civil society engagement or material activism
10 Resistance to change within government agencies
11 Not significant awareness and support across industry and society.

GHG emissions through the common but differentiated responsibilities (CBDR) provisions.

This experiment is illuminating in that it draws attention to the direct links between the defined objectives and the outcomes. In the instance of EM, the objective is the decoupling of economic development and environmental protection whereas the UNFCCC seeks the containment of global warming, economic growth and social development, and the EU and UK add energy security. Accordingly, the expectations of the contribution of the theoretical framework can only be to that common aspect of the defined objectives.

CONCLUSION

The theoretical framework presented here has provided a set of conditions to contextualize the application of EM in practical policy contexts. To that end, and as established in the exercise, an operationalized EM serves to identify where there are gaps in the suite of policies and programmes, that may inhibit the ability to achieve the desired ecological outcome, recognizing there may be other factors that could also create such a barrier, but are outside the scope of the theoretical framework.

UNFCCC's Kyoto Protocol and the INDCs under the 2015 Paris Agreement provide for ratifying nations to commit to specific targets and timeframes concerning emission reductions and many member states, particularly in developed economies, have framed domestic policies with these targets as outcomes. While the international agreement and the domestic policies examined in the experiment can be described as models of optimum EM, the multiple objectives of those agreements and policies may mitigate against and complicate progress towards the achievement of the desired ecological outcome. In the cases of India and China, the constraints imposed on climate policy delivered a sub-optimum result. Consequently, it is worth recognizing that there is tension between 'achieving the achievable' in environmental policy, and subordinating the required ecological outcomes to external policy imperatives. The requirement that international agreements and regional/national level implementation of climate policies should facilitate economic growth overlooks concerns about the finite nature of available resources, and subverts the requirement to change methods of production and consumption. The principle of common but differentiated responsibilities, while honourable and possibly even appropriate in 1992 when the Convention was adopted, provided an exemption for countries that are now major economies as well as the largest GHG emitters. The Paris Agreement has partially addressed this historical legacy, but CBDR may well continue to cast a shadow over the implementation of optimum EM models for developing countries.

NOTE

1. Clean production is production without (unrecycled) waste or emissions (Janicke et al., 1997).

2. The role of employers' organizations and trade unions in the climate policy process

INTRODUCTION

The contention that has guided this research is that employers' organizations and trade unions, as representatives of the actors in the workplace and as the holders of a unique body of relevant knowledge about the impact on the labour market, are integral to the development and implementation of effective climate change policy. This chapter establishes the impact of climate change policy on the labour market and examines the role of employers' organizations and trade unions in the process of related policy development. It explores the labour market impacts of climate change policy, the role of employers' organizations and trade unions and the role of civil society organizations.

THE LABOUR MARKET IMPACTS OF CLIMATE CHANGE POLICY

The link between climate change and the labour market was formally acknowledged when the Heads of State and Government in 2009 adopted the recommendations of the UNFCCC Adaptation Working Group that climate change agreements extend the social dimension of policy and provide for a just transition and decent work (UNFCCC AWG LCA, 2009). A just transition is defined as the recognition of workers' rights, decent work, social protection and social dialogue (Worldwatch Institute, 2008). The four tenets of decent work as articulated by the ILO are creating good jobs, guaranteeing respect for workers and the recognition of their rights, extending social protection and promoting social dialogue (ILO, 2011a). The importance of these issues was identified as a result of studies undertaken by the ILO, ETUC and the UK government that established the direct relationship between climate change and the world of work.

The emergence of labour issues in formal climate agreements reflects the growing acceptance that climate change has impacts across the broader economy and society. While Nicholas Stern (2007) and Ross Garnaut (2008) established the bridge to understanding the economic impacts of climate change on a nation's economy, the research for the ILO (Worldwatch Institute 2008), the UK government (GHK Consulting 2007) and the ETUC (2007) was instrumental in creating awareness of the employment and workplace impacts of climate change.[1]

The Worldwatch Institute's (2008) report for the ILO encapsulates the findings that are common across research on the subject and holds true today. Due to the impact of climate change on public policy and the economy, there will be a consequential impact on the labour market and some jobs will be lost, some jobs created and some jobs changed. On balance, there will be modest net growth in employment and all sectors of industry will be affected. The research contended that governments must have policies in place to manage the changes.

The research by the ILO International Institute of Labour Studies (ILO, 2012), OECD (2016) and others adds that outcomes for employment and incomes are largely determined by the policy instruments and the institutions that implement them, rather than being an inherent part of a shift to a greener economy. The research also finds that policy needs to be tailored to the circumstances of the country and that there is no single policy template that can be applied in all circumstances.

The OECD *Green growth strategy synthesis report* (OECD, 2011) and subsequent OECD research discusses the employment and distributional aspects of green economy policy, noting that occupational churn will create challenges for some and opportunities for others. The OECD contends that labour and skills policies are important so that workers and companies are able to adapt quickly to changes and to seize new opportunities.

ILO (2012), Eurofound (2011b) and OECD (2013a) research found that the quality of jobs is enhanced when labour market institutions, the education and training system and the industry communicate regularly, as well as when social partners and, notably, trade unions are involved in the process of transformation. Employers' organizations and trade unions play an important role in supporting local jobs and training as well as facilitating industry information valuable for the local labour market (Miranda and Larcombe, 2012).

The labour market in higher GHG-emitting sectors in domestic economies is directly affected through regulatory and market measures that are aimed at reducing dependence on fossil fuels and GHG emissions. Buildings are a major contributor to the emissions problem as they are traditionally large and inefficient consumers of energy and producers of greenhouse gas emissions. The 2011 World Economic Forum report *A profitable resource-efficient future* states that buildings use 40 per cent of the world's energy and emit 40 per cent of the world's carbon footprint (WEF, 2011). However, buildings are also part of the solution, as nearly half of all energy consumed in buildings can be avoided with the use of more energy-efficient systems and equipment (WEF, 2011). Reports published in 2011 analysed the construction sector labour market (Gleeson et al., 2011; ILO, 2011d; OECD, 2011) and predicted that 'green' construction meant that some industry occupations would evolve while, in other cases, converting to a green construction team would add occupational profiles and new occupations. The ILO (2011c) forecasts the emergence of additional soft skills functions such as assurance, financing, research, education and policymaking. Gleeson et al. (2011) asserted that green construction teams would require competent emissions assessors, project managers, assessors, appraisers, skilled labour and auditors. An Austrian study reported the additional competencies required of a new green plumber tradesperson, beyond the traditional specialist technical skills, included customer orientation, the ability to communicate, decide, consult and sell, as well as planning competencies, and a high level of independence and global thinking (Friedl-Schafferhans, 2011).

The EU member states' strategies to transition the construction sector reflected the needs of their different domestic situations while remaining within the framework of the EU 2030 Climate and Energy Framework (EC, 2014). The French government's Energy Transition for Green Growth Act (2015) advances the platform created by the Grenelle Plan-Bâtiment to deliver renovation of 500,000 housing units per year and create 75,000 new jobs. All new buildings were to be low power and to be 'positive energy' by 2020 (ADEME, 2011).[2] A UK flagship initiative was the Green Deal, intended to reduce emissions from domestic and commercial dwellings by implementing energy-efficiency measures and retrofitting (DECC, 2011d). While the Green Deal was a well-resourced programme, there were concerns that its proposals for labour market reform were insufficient to meet the demand for labour in the number and with the skills required to carry out the work (Gleeson et al., 2011). In the end, the programme failed to engage with consumers and finance was withdrawn in 2015 without any successor programme being announced. The German government undertook the renovation of existing housing stock at the rate of 300,000 apartments per year (Syndex, 2011). Their objectives are for buildings to be virtually carbon neutral by 2050. The government also sought to create and preserve 200,000 jobs, reduce emissions by 2 million tonnes per year, drive down energy bills, reduce the state debt by a minimum of €4 billion and reduce the country's dependence on fossil fuels (Syndex, 2011).

In the transport sector, the shift to a sustainable mobility system is expected to create jobs and enhance social equity. It is believed new jobs will be created in the transition, which, over time, will exceed the loss of 'brown' economy[3] jobs. Transport is fundamental to functioning of labour markets and economies. The 2011 European Commission White Paper *Roadmap to a Single European Transport Area* states that in 2011, the transport sector employed around 10 million people (60 per cent in road, 30 per cent in rail and public transport and 10 per cent in air) and accounted for 5 per cent of GDP (EC, 2011c).

However, while there is consensus that climate change is a major threat and will have an impact on employment, there is intense disagreement about how individuals and organizations should respond and what action should be taken (Miranda and Larcombe, 2012). While the debate continues about the appropriate policy options, in some sectors the options are limited. In transport, where motor vehicles are the major contributor to greenhouse gas emissions, congestion and pollution problems, the only option to reduce the climate impact entails diverting public policy and investment away from roads and trucks to public transport modes. The consequence is likely to be large-scale shifts of employment within bands across firms in the sector and will require retraining, skill upgrading and career transitioning.

As certain regions are highly dependent on motor vehicle factories and related employment, it is considered that the success of any strategy is enhanced by the extent to which the strategy is informed by social dialogue. Bringing employers' organizations and trade unions to advise during the transition phase augments the responsiveness of the stakeholders and triggers green transformation on a larger scale (ILO, 2012; Eurofound, 2011a).

UNFCCC climate agreements are drafted around the acknowledgement that no single approach can apply to all circumstances (UNFCCC, 2013). In the EU, policy remains the responsibility of the state and EU members' strategies are tailored to domestic economic and social situations. The French government has committed to the EU's 2030 Climate and Energy Framework and has also resolved to ratify the 2015 UNFCCC Paris Agreement (UNFCCC, 2015). In 2007, the French government hosted a series of domestic stakeholder consultations. The government adopted the programmes that emerged from these consultations into two outcomes: Grenelle 1 and 2 (Euractiv, 2010), a process that has continued and has seen the policy constantly evolve. In 2012, the government hosted the first environmental conference, which facilitated dialogue between all stakeholders and with territorial authorities and Members of Parliament. This was followed in 2013 by the National Debate on Energy Transition

(Ministry of Ecology, Sustainable Development and Energy, 2013). The government's current strategy is the 'ecological transition towards sustainable development. A new strategy for 2015–2020' (Ministry of Ecology, Sustainable Development and Energy, 2015). This entails shifting towards a new economic and social model that is underpinned by two components, social and societal innovation accompanied by new modes of governance, production, consumption and cultural collective references; and technological innovation and research and development in the area of organization and industrial processes. Its labour market plan remains the *Green growth mobilization plan* (Ministère de l'Écologie, 2010), which addresses jobs growth and is coordinated across the education, business and regional development portfolios. The Energy Transition for Green Growth Act (2015) frames the plan's current strategies.

The German government met its commitments under the Kyoto Protocol by 2008 and has committed to a further reduction in emissions of 40 per cent by 2020 and 80 per cent by 2050 (Wilke, 2011). Its labour market arrangements provide for greening of the occupational profiles and formal vocational training which have evolved naturally with the progressive greening of the economy (Strietska-Ilina et al., 2011).

THE ROLE OF EMPLOYERS' ORGANIZATIONS AND TRADE UNIONS

Defining Key Terms

The following section explains the role of employers' organizations and trade unions in the policy development process. But before their roles can be examined, it is necessary to understand the terms 'employers' organizations' and 'trade unions'. *The Macquarie Dictionary of Australian Politics* (Macquarie, 2013) defines employers' organizations simply and adequately as collective organizations of employers. *The Blackwell Dictionary of Political Science* (Blackwell, 1999a) is almost as brief, describing them as organizations of employers combined to defend their common interests. Trade unions were equally simple to define. Blackwell (1999b) defines them as collective organizations of employees, formed to safeguard the terms and working conditions of their members while *The Cambridge Dictionary of Sociology* (Cambridge Dictionary, 2013) defines them as formal organizations of workers that seek to represent the interests of their members through collective organization and activity, offsetting the weakness of individual employees compared with the power of employers and managers. The ILO, OECD and the EU have not published definitions, although their application of the terms is consistent with the above. For the purpose of the examination that follows, Blackwell's definitions are accepted; that is, employers' organizations are collective organizations of employers formed to defend a common interest, and trade unions are understood to be collective organizations of employees formed to safeguard the terms and conditions of their members.

Other related terms relevant in this research and sometimes used interchangeably are business and industry associations, civil society organizations (CSOs) and non-state actors (NSAs). Employers' organizations and trade unions are both CSOs and NSAs but the converse does not necessarily apply.

Business and industry associations like employers' organizations are membership organizations. Membership is generally voluntary and subject only to eligibility for membership, agreement to observe the rules of the association and to pay the prescribed fees. Business and industry associations are often mandated to represent their members on all issues related to their members' interests, which may include their responsibilities as employers. On the other hand, businesses may seek to separate responsibilities and confer authority with a dedicated employers' organization and a business or industry association for commercial matters.

For the purposes of this discussion civil society is the collective term for CSOs. The World Bank refers to civil society as 'the wide array of non-governmental and not-for-profit organizations that have a presence in public life, expressing the interests and values of their members or others based on ethical, cultural, political, scientific, religious or philanthropic considerations' (World Bank, 2013, p. 1). 'Non-state actor' is a term used to refer to non-governmental organizations, which may be civil society organizations, transnational corporations and intergovernmental organizations (Weiss et al., 2013; Joey, 2015). The Oxford Dictionary definition of an NSA is an individual or organization that has significant political influence but is not allied to any particular country or state (Oxford Dictionary, 2015). Another more descriptive definition is non-governmental organizations and other civil society groups, and also other non-state entities having some impact upon international relations, including multinational corporations, terrorist groups, and organized criminal gangs (Australian Legal Dictionary, 2015). In the context of this research, employers' organizations, business and industry associations, and trade unions are also CSOs, where CSOs are representative organizations, NSAs are any of CSOs, individual persons, corporations and non-governmental organizations. In essence, while CSOs are NSAs, not all NSAs can be CSOs.

It also is important to consider the different objectives of the organizations to which these definitions apply. Trade unions have been described as activists on workplace issues and employers' organizations as associations of like interests. Plowman (1978) contends that employers' organizations are reactive by nature, which has allowed unions to effectively dictate the industrial agenda, that it is only the actions of unions that bring employers together and at times when there is no union activity, employer groups lose interest and dissipate. He does not mean that the organizations are scaled back to the point of non-existence, but that the interests of the organization and its members shift to other issues of relevance to their business and industry needs which they believe benefit from a collective presence. The mandate shifts from that of representative of their interests as employers to that of representative of the business interests. For organizations with a dedicated employment mandate, their financial viability and their effectiveness as an activist with government is influenced by their industrial relations responsibility and the incidence of union activity (Croucher et al., 2006; Traxler, 2010). State-based support for multi-employer bargaining has the effect of delivering to employers' organizations a grip on the labour market that prompts governments to involve them in public policy which then pulls in trade unions, given the practice of involving organized interests in public policy according to the principle of class parity (Traxler, 2010).

Blackwell's (1999b) description of trade unions as collective organizations of employees formed to safeguard the terms and conditions of workers is commonly accepted (Hyman, 2001; Kelly and Heery, 1994). The representative role of trade unions brings further responsibilities, as agencies of class in the labour management conflict, and as functionaries within the social framework, which they aspire to change. Hyman (2001) observes that in times of change and challenge for union movements, a reorientation may occur between their responsibilities to the market, society and class. The conflict that arises in the process of reorientation due to ideological differences amongst the membership and the factions creates further problems for the trade union in periods of declining membership. Survival requires ideological differences be put aside in order to conserve, mobilize and target union resources on the immediate priority issues (Kelly and Heery, 1994).

The objectives of employers' organizations and trade unions overlap only with regards to workers' wages and conditions and union workplace activism. The dichotomy is demonstrated at the international climate negotiations in which trade unions, for example, have successfully advocated the inclusion of decent work and a just transition in the formal texts (UNFCCC, 2010; UNCSD, 2012a; UNFCCC, 2015) and have done so unopposed by business associations which have focused their advocacy on trade and energy (ICC, 2012).

International Context

The peak international organizations for business and employers are the International Chamber of Commerce (ICC) and the International Organisation of Employers (IOE) respectively. The ICC has consistently advocated that international climate agreements must be considerate of the consequential impact on business and that business is an essential link in the value chain that will deliver the environmental outcome (ICC, 2010, 2015a). The advocacy focus of its constituents at the UNFCCC COPs has been trade and energy, while at the Rio + 20 Conference it was the capacity of trade and investment to help create sustainable jobs and to make a greener and fairer economy happen (ICC, 2012). Neither the ICC nor the IOE have responded to the labour and employment proposals for decent work and a just transition that emerged firstly in the 2010 UNFCCC Cancun Agreement (UNFCCC, 2010) and then the Rio + 20 Conference outcomes and in trade union advocacy to the Conferences.

The peak organization for workers is the International Trade Union Confederation (ITUC). The ITUC and its predecessor organizations the International Confederation of Free Trade Unions (ICFTU) and the World Confederation of Labour (WCL) have attended the international climate and sustainability conferences since the 1992 Rio Summit. Their policy position has evolved, albeit around the same themes of worker equity and a just transition. At the 2010 Cancun UNFCCC COP the ITUC catch-cry was 'What do we want in Cancun? A Just Transition NOW' (ITUC, 2011). Other rallying cries, echoing environmental NGOs, include 'there are no jobs on a dead planet' (Burrow, 2015). The ITUC has been the primary and possibly only non-governmental advocate for the inclusion of labour and employment issues in COP agreements. The ITUC has proposed a global financial transaction tax and a carbon tax as means of financing the sustainability agenda, for funding the transition of workers and protecting their retirement entitlements (ITUC, 2012b; Burrow, 2015).

The formal roles of the ICC and the ITUC in UN sustainability and climate change activities are as the central contacts and administrators of the Business and Industry, and Trade Union and Workers Major Groups.[4] Although these two major groups have many issues of common interest in their pursuit of a favourable commercial and industrial environment for their members, they rarely interact at COPs. The Trade Union and Workers Group has taken up advocacy of the decent work agenda while the Business and Industry Group does not operate as an advocate but as an agency through which the sectoral or domestic interests of its members can individually participate in proceedings, acting as a forum for exchange, gathering and disseminating information. Prima facie, given the number of trade union representatives at climate conferences (historically less than 1 per cent), the impact of organized labour on determining the outcomes of proceedings appears greater than the sum of its parts, although studies show that environmental, academic, business and energy interests appear to have a greater level of influence (Cabré, 2011). Business and industry is not motivated in a similar fashion, and does not have a mandate or a formal constituency. In practical terms, it is of no matter to the ICC how the sectoral interests use the opportunity, and neither is it their interest to measure the outcome, but rather whether they are able to optimize the opportunity for the delegates. That said, the business representation has been high level, and so it is reasonable to assume that the advocates at the negotiations are credible and supported by market intelligence.

European Context

The 1993 Maastricht Treaty, the treaty that forms the EU, provides for a highly competitive social market economy aiming at full employment, social progress and improvement of the environment (Europa, 2012b). It requires that employers' organizations and worker representatives are consulted on matters of social policy, social and economic cohesion, environment, health and consumer protection. In European regulation, employers' organizations and trade unions are afforded the title of social partners and are viewed as playing an important role in the economy as a whole and the labour market in particular.

The European Commission's reports *Industrial relations in Europe* (EC, 2011b; 2013; 2015) find that in many EU member states, the social partners are involved with low carbon economy issues from the stage of policy formulation. The European Commission looks to the social partners to create consensus for policies across industry and society. It has confidence in their ability and leadership, expressing the view that 'a shared analysis of employment opportunities and challenges by social partners can contribute greatly to a well-managed and socially just transition' (EC, 2011b, p. 153). The reports observe that employers' organizations and trade unions have been positive contributors to the development of public policy and the implementation of adaptation and mitigation strategies, interacting with the nation-state, the actors in the economy and others in civil society to facilitate the transition. That said, and while reports by some agencies of the European Commission (Eurofound, 2011b; Syndex, 2011) contend that the most successful adjustments to the labour market in the transition to a low carbon economy have taken place when social partners were involved in the process of transformation, in the majority of member states the low carbon economy and its employment consequences appear to remain marginal items on the agenda of social partners (EC, 2011b, p. 161; EC, 2013; 2015).

The ITUC advocates that climate change should be a matter for collective bargaining where issues of interest to organizations' representative of employers and workers are subject to negotiation and inclusion in collective bargaining agreements (Morris, 2010). Collective bargaining is usually associated with the industrial relations process of negotiating wages and conditions of employment but, in the European statutory jurisdiction, the collective bargaining process is also being used as a regulatory tool with the collective agreement as the regulatory instrument (de Boer et al., 2005; Smismans, 2008). The IOE's view is that climate change may be a matter for social dialogue but not collective bargaining, arguing that as an issue for labour and management, climate change does not add a condition to existing labour agreements and that issues arising from climate change where relevant to the business should be addressed through the normal consultation process (IOE, 2009). Social dialogue is defined as all types of formal dialogue involving discussions, consultations, negotiations and joint actions undertaken by employer and worker representatives (Eurofound, 2011c). In Europe, social dialogue at company level on issues such as energy efficiency and climate change is slowly spreading but it is not common practice (EC, 2011a; 2013; 2015).

UK Context

Employers' organizations and trade unions have a degree of recognition across the community as the representative organizations for their members and, as such, are often invited to contribute to the process of public policy development and its implementation. Their advocacy on climate change reflects their different philosophical perspectives – employers' organizations are concerned with ensuring companies remain competitive in the transition to a green economy (Eurofound, 2011b; BusinessEurope, 2015a), while trade unions insist on a just transition that includes social dialogue, skill adaptation and investment in green jobs (EC, 2011b; Burrow, 2015).

The largest employer organization in the UK, the Confederation of British Industry (CBI), declared its commitment to action on climate change early and openly, although it has not adopted a formal policy position. In 2007 the CBI published the report *Climate change: Everyone's business* (CBI, 2007). In the absence of a formally stated policy, this report became the CBI's manifesto on climate change. The CBI formed a specialist advisory committee of the major UK businesses and resourced a secretariat that was active in public discussion and provided guidance to members through information and awareness. The key messages of the 2007 CBI report were the declaration of support for the government's emission reduction and energy-efficiency strategies and that the UK had a unique opportunity to prosper by taking a lead in the development of low carbon technologies and services.

The CBI's advisory committee, the Energy and Climate Change Board, describes itself as a group of business leaders committed to tackling the UK's triple challenges of energy security, affordability and de-carbonization. It was particularly active in the lead-up to the 2015 Paris COP, releasing the report *Setting the bar: Energy and climate change priorities for the government* (CBI, 2015a), which offered advice on the business requirements of UK, EU and international policy.

While it is the largest membership organization representing business in the UK, the CBI is not a formal peak organization. Its membership is 1,500 direct corporate members and 140 industry associations. Its association members advocate on their own behalf except on issue-specific occasions when they agree that CBI should be the advocate. On climate change, the CBI is not briefed to act on their behalf. Bailey and Rupp (2006) provide a pithy assessment of business representation on the UK, observing that the organizations experienced poor credibility with government because there were so many of them and they lacked a strong peak organization. As a consequence, the UK government often chose to deal directly with the leading companies. In that regard, the CBI is strategically positioned through the membership of its Energy and Climate Change Board, who are the leaders of industry and the business contacts government requires to support and facilitate policy implementation.

The peak organization of trade unions in the UK is the Trades Union Congress (TUC). The TUC has focused its domestic efforts on the Green Workplaces initiative, a programme it developed to bring together both workers and management to secure energy savings and reduce the environmental impact of their workplace. The TUC provides training for union sustainability representatives on the technical issues related to climate change adaptation and mitigation and negotiating with employers. The TUC (2010) initiated a campaign for union representatives to be afforded the right to time off during working hours to promote sustainable workplace practices, to receive training and to inspect workplaces for energy efficiency. However, while it maintains this policy position, there is no indication that it has any traction with government or employers.

The continued decline of union density (membership) and the decentralization of collective bargaining in the UK have impacted unions' influence as a representative force and accordingly their ability to influence governments and policy development. While UK trade union penetration for the public sector is 54.8 per cent in 2015, in the private sector is only 13.9 per cent (BIS, 2016). Climate change presented an opportunity for unions to reinvent themselves and the Green Workplaces initiative is part of the unions' renewal strategy (EC, 2011b). However, since the release of Green Workplaces the TUC has not maintained the effort and the opportunity may have passed.

The influence of employers' organizations and trade unions over domestic policy and the contribution to the country's targets for GHG emissions will vary in each country. This discussion of the UK context establishes the scope and ability of these organizations to influence policy in Europe, the constraints imposed by the market and that their authority is directly related to their membership strength; these factors are common across all employers' organizations and trade unions.

THE ROLE OF CIVIL SOCIETY IN THE CLIMATE CHANGE POLICY ARENA

Overview

Employers' organizations and trade unions are active and well-resourced advocates who also engage in networks whose policy and advocacy objective is complementary. Civil society as stakeholders in climate change are extensive and dedicated and often highly regarded. This book adopts ecological modernization as the theoretical framework and while civil society is an integral element of an EM model, it also examines whether employer organizations and trade unions should be recognized separately or whether EM in its current form appropriately accommodates the influences over the ecological outcome.

The literature addressing the role of civil society in the climate change debate is extensive. Academic and institutional interest in the role and activities of civil society increased following the declarations from the 2012 Rio + 20 Conference and the subsequent 2012 Doha COP that applauded the contributions of civil society to the debate. Other agencies such as the G20 have since followed suit and the G20 Leaders' Summit, held in St Petersburg in September 2013, permitted civil society the opportunity for input to the formal deliberations (C20, 2012).

Defining Civil Society

Civil society is the collective term for civil society organizations (CSOs). The World Bank refers to civil society as:

> the wide array of non-governmental and not-for-profit organizations that have a presence in public life, expressing the interests and values of their members or others, based on ethical, cultural, political, scientific, religious or philanthropic considerations. Civil Society Organizations (CSOs) therefore refer to a wide of array of organizations: community groups, non-governmental organizations (NGOs), labour unions, indigenous groups, charitable organizations, faith-based organizations, professional associations, and foundations. (World Bank, 2013, p. 1)

The OECD and UN do not explicitly define civil society but adopt the same broad comprehensive scope as the World Bank.

While EM theorists generally discuss civil society activists in the context of environmental organizations (for example, Mol, 2008 and Gonzales, 2009), in practice the scope and interests of civil society are much broader. The UNFCCC uses the term 'observer' to describe those non-member state organizations that are permitted to attend meetings and conferences. These organizations represent a broad spectrum of interests embracing representatives from business and industry, environmental groups, farming and agriculture, indigenous populations, local governments and municipal authorities, research and academic institutes, labour unions, women and gender and youth groups (UNFCCC, 2011b).

CSOs provide informed advocacy, bring expertise related to their specific interests and provide information that is often otherwise not commonly available. Governments and regulators may seek CSOs' contributions to public debate and public service, because of their specialist knowledge and their relationship with stakeholder groups. Government advisory boards may have civil society representatives[5] and CSOs may be invited to assist with the delivery of government services[6] and may act as a regulator in their sphere of interest.[7]

CSOs can act individually or may form coalitions of organizations with like interests to increase the effectiveness of advocacy. The European Trade Union Confederation (ETUC), with the CSOs European Environment Bureau and Social Platform, campaigns for a social and sustainable Europe. These three organizations make annual recommendations to the EU about maintaining a balanced approach to the three sustainability pillars of economic, social and environmental action (ETUC, 2007). It joined with the Spring Alliance of unions and civil society to call for concrete actions from the Rio + 20 Summit (ETUC, 2012) and with BusinessEurope and others to form the European Alliance for Apprentices (ETUC, 2016). On matters related to climate change, BusinessEurope acts with the CSO Alliance for a Competitive European Industry (ACEI) (EC, 2011b). The ACEI's objective is to promote the competitiveness of European industry on a global scale. On climate change, it takes a fully integrated approach to industrial policy by carefully balancing essential climate, energy and competitiveness factors (ACEI, 2013).

The fabric that makes up civil society is diverse. The CSO CARE International is a humanitarian organization fighting poverty: it is a service provider, activist and advocate (CARE Australia, 2010). The CSO Global Humanitarian Forum, which wound up in 2012, was formed to address humanitarian challenges including research into the human impacts of climate change. It leveraged the profile of its presiding chair, the retired UN Secretary-General Kofi Annan, to create a point of difference that enhanced its authority (Dalber Global Development Advisors, 2009).

CSOs have been prominent in shaping public policy on the environment and climate change. In the UK, the Friends of the Earth (FoE) campaigned under the 'Big Ask' banner for a legislative commitment to GHG emission reduction. The World Wide Fund for Nature (WWF) campaigned for UK energy companies to commit to low emissions technology (WWF, 2003) and the Carbon Disclosure Project (CDP) sought the voluntary inclusion of socially responsible investment criteria in investment decisions (Pfeifer and Sullivan, 2008). Until the 'Big Ask' campaign, the UK government's climate strategy was a blend of hard policy through tools such as the voluntary Climate Change Levy introduced in 2001, the Emissions Trading Scheme (UK ETS) in 2002 and soft policy option of corporate social responsibility initiatives. The 'Big Ask' campaign convinced a cross party group of Members of Parliament to table its draft Climate Change Bill in parliament. The Bill provided a legal framework to manage future emissions (Hall and Taplin, 2007) and became the foundation of the Bill that was finally passed into law on 28 October 2008 (FoE, 2008; Lock, 2006).

Although EU and member states boast the engagement of civil society in decision-making, in practice the consultation is not always undertaken with a transparent objective. Some observers contend the consultation is a process used by the government agencies only to advise interest groups how development is to be undertaken and how they will be affected, to confirm established decisions or to legitimize the policy (Braun, 2010; Blackstock et al., 2006; Scheer and Hoppner, 2010). This means that civil society efforts to contribute to ensure informed policy are frustrated and may have little or no influence over decision-making although recent EU directives have addressed this abuse of process to some extent (Braun, 2010).

Business and industry associations, employers' organizations and trade unions are CSOs but differ from many in that there is awareness across society of what they do and why. This awareness is the product of generations of activism and many in the community have come into contact with them through their engagement in industry or workplace issues, either as management or labour. They also have wide coverage with affiliations that extend across jurisdictions from the international to regional, national, sectoral and the workplace. There are few other categories of organization within civil society that share these credentials or penetration.

Civil Society within the United Nations

Civil society has a significant presence at international climate change negotiation events and is welcomed as a valuable source of specialist knowledge to negotiators (Brazil CSD, 2012). The UN system is generally open to contributions and participation by civil society. At the Rio + 20 Summit, CSOs were invited to make written submissions for inclusion in the compilation document, *The Zero Draft* (UNCSD, 2012b), and were active advocates during the pre-conference negotiations.

The Rio + 20 *Future we want – outcome document* offered formal recognition of the contribution by civil society declaring that 'We, the Heads of State and Government . . . with the full participation of civil society, renew our commitment to sustainable development' (UNCSD, 2012a, Article 1). The statement was somewhat misleading, as civil society was not a negotiating party and neither was its agreement sought.

The actions by the UN and member states to engage civil society organizations reflects the perception that civil society is ready, willing and able to take on the responsibility that comes with a formalized role in the negotiations. However, it remains that the ability of civil society to act cohesively and collectively has not been tested. Civil society is comprised of many organizations with unique and specialized interests with a way of doing business that is not a natural fit with the formal UN requirements for agreed and consistent process and modes of behaviour. In research on the behaviour and the tactics of CSOs in UN-based policy negotiations it has been contended that CSOs push the boundaries in negotiations using tactics that are not always consistent with the UN's standard practices. During the course of deliberations, CSOs work strategically to maintain their independent presence (Eastwood, 2011). Against that background, it could be a challenge firstly to achieve agreement among the unique and specialized interests that are CSOs to be part of a discrete homogenous and representative entity, and secondly to achieve participation in a collegiate model that would require them to cede authority to an elected representative(s) of the civil society community.

Different schools of thought have emerged in the debate about what management model is most suitable for the possible representation of the collective interests of CSOs. The Secretary-General of the World Federation of United Nations Associations, Bonian Golmohammadi, offers the view that civil society needs to adapt to any requirements of the UN for participation because the UN is the sole organization capable of shouldering the sustainability responsibility (Golmohammadi, 2012). He proposes a participation model similar to the ILO in which business and worker representatives sit at the table with governments and participate in the debate with formal rights to speak but only governments vote. At the other end of the scale is the existing model of activism typical of independent CSOs that would push the boundaries of accepted practice and protect their independence (Eastwood, 2011).

The Chair of the UN High Level Political Forum on Sustainable Development (UNHLPF), that in 2012 succeeded the UN Commission on Sustainability, invited CSOs through the UN Major Groups to engage in a process and negotiations that would determine the UNHLPF modalities and format. In response, the Major Groups stated that in their view the full and effective participation of civil society could only be achieved if, as a minimum, procedures were put in place that ensured public disclosure, access to information and public participation. They explained that in practice this would mean Major Groups, and by implication their participating CSOs having access to all documents and drafts in a timely manner; being permitted to comment on draft reports and receiving an official response; having time allocated for dedicated dialogue; having access to regular meetings with Bureau members; and having access to all meetings at all levels with speaking rights (UNHLPF, 2013).

The meeting of the UNFCCC APA (Ad Hoc Working Group on the Paris Agreement), held in Bonn in May 2016, added to these earlier considerations, commencing discussion on improving observer engagement. The meeting decided to convene an in-session workshop at the meeting SBI 46 held in May 2017 'on opportunities to further enhance the effective engagement of non-party stakeholders with a view to strengthening implementation of the provisions of decision 1/CP.21 Paris Agreement' (UNFCCC, 2015). The UNFCCC Secretariat of the Convention went so far as to prepare a report to inform discussions within the workshop.

The binding nature of any rules for civil society engagement is certain to be tested by the many CSOs to ensure the right to advocate their particular cause is not compromised. The Major Groups created within the UN system are not official or mandated representative bodies and do not have authority beyond being a contact point and interface with that system. At the climate COPs and events such as the Rio + 20 Conference, CSOs are not bound to the organization that leads the Major Group, its philosophy or ideology. The exception may be the trade union Major Group, but this is more a reflection of the unanimity of their purpose and their collegiate structure and therefore the advocacy of an agreed manifesto.

The attention to the economic and social issues and the perceived needs of civil society had an effect on the efficiency of the climate change negotiating process. After the 2012 Doha COP, BusinessEurope reported that progress on negotiating issues at the conference was slow and the process was inefficient (BusinessEurope, 2012). The Climate Group wrote of the weak outcome and stated that some tough issues were simply ignored (Ryan, 2012). These statements parallel the situation that arose from the 2011 Durban COP 17, which was criticized because of the deferral of decisions to the following COP in Doha and for otherwise making no decisions that helped bridge the ambition gap (Climate Action Network, 2012). This lack of material progress fostered little confidence that the negotiators would be able to reach agreement on a successor to the Kyoto Protocol (Rajamani, 2012), a concern that was not bridged until the 2015 COP that delivered the Paris Agreement and the Lima Paris Action Agenda.

CONCLUSION

The decision to include the employment and social dimensions of climate change in the international climate agreements has created an unexpected paradox. While the negotiating parties to the agreements were debating the process to formalize the role for civil society, civil society was experiencing barriers to effective advocacy. The Bali Action Plan (UNFCCC, 2007) operationalized the employment and social dimensions of adaptation and mitigation policy, measures seen as equitable and necessary but, with hindsight, possibly not achievable under the model in place, as this model was established with the single objective of managing the pace of change in the climate. There are many outstanding issues that support the need for a review of the governance and institutional arrangements to manage climate change and sustainability and to ensure that CSOs can contribute effectively and efficiently to the debate.

NOTES

1. Also see OECD (2016), Taylor et al. (2016), Rademackers and van der Laar (2014).
2. It has not been possible to establish the performance against these targets.
3. A polluting economy (Thomas, 2013).
4. Since the first UN Conference on Environment and Development in 1992 – the Earth Summit – it was recognized that sustainable development could not be achieved by governments alone. This notion is reflected emphatically in the landmark outcome document of that Summit, 'Agenda 21'. Section 3 underscores the criticality of harnessing expertise and capacity from all sectors of society and all types of people: consumers, workers, business owners, farmers, students, teachers, researchers, activists, indigenous peoples

and other communities of interest. Agenda 21 formalized this concept by recognizing nine sectors of society as the main channels through which citizens could organize and participate in international efforts to achieve sustainable development through the UN. These nine sectors are officially known as Major Groups (UN SDKP, 2016). As noted previously, in UNFCCC they are referred to as BINGOs and TUNGOs.

5. For example, the Australian Government Climate Change Advisory Committee (Combet, 2010).
6. For example, in the UK in 2012 CSOs were used to accredit and up skill installers, and communicate to industry and the consumer the Green Deal initiative (ECA UK, 2011).
7. UK ETS allocation of permits to industry (Dresner et al., 2006).

3. Climate policy in context I: countries within the EU

The research for this book establishes the role of employers' organizations and trade unions in the climate change policy development process and was based on a qualitative study of the peak representative organizations of employers and workers in the policy process in the international, regional and domestic jurisdictional contexts. International climate change agreements frame the domestic policy process and the European regional policy and programmes are a mature and comprehensive suite of actions, taken up across the 28 member states and are at the forefront of climate change action globally. The eight countries chosen include three of the major industrialized economies in Europe, UK, France and Germany that are accordingly a major influence over European policy and make major contributions to the performance against targets. In selecting the other countries, the common threads that would impact the policy process were identified as the parliamentary and legal process, cultural influences and language. Australia, Canada, Singapore, India and Kenya shared these threads through their historical ties to the British Commonwealth.

The outcomes from the intergovernmental negotiations under the UN Framework Convention on Climate Change (UNFCCC) reflect the terms of the commitment by the Parties to the Convention and the terms under which they will act in their domestic situation. At the 2015 Paris COP the Parties agreed to a range of legally binding conditions, voluntary commitments through their individually determined contributions, commitment to provide finance for mitigation and adaptation initiatives and the role of non-state actors (UNFCCC LPAA, 2016). Previously, the developed countries committed through Annex 1 of the Kyoto Protocol.

The European Union is a party to the Kyoto Protocol and has committed to ratifying the Paris Agreement. The EU acts in these matters on behalf of its 28 member states, which are bound by the EU's commitment but can implement in accord with and as appropriate to their domestic circumstances. The EU has adopted the 2030 Climate and Energy Framework which sets three key targets for the year 2030 (Europa, 2016):

- At least a 40 per cent cut in greenhouse gas emissions (from 1990 levels);
- At least a 27 per cent share for renewable energy;
- At least 27 per cent improvement in energy efficiency.

The framework was adopted by EU leaders in October 2014. It builds on the *2020 climate and energy package* (Europa, 2010a). It is also in line with the longer term perspective set out in the *Roadmap for moving to a competitive low carbon economy in 2050*, the *Energy roadmap 2050* and the *Transport white paper*. The EU provides a mature model of sustainability and climate change policy development, effective and considered advocacy in related international and intergovernmental fora and effective management and accountability of the member state activities, industry and the market, civil society and community awareness.

The UK, France and Germany are the largest economies in Europe. Each met their commitments under the Kyoto Protocol, have adopted the EU 2030 Climate and Energy Framework and have in place strategies to advance their commitment to a domestic sustainability agenda (Europa, 2016). The UK 2050 Pathway Analysis (DECC, 2010) and the Carbon Plan (DECC, 2011a), are implemented through the Climate Change Act 2008 and the Energy Act 2008. As such, they represent the UK's vision and plans for reducing emissions, climate management and energy efficiency. In 2010, the French government adopted the programmes that emerged from its government-hosted stakeholder consultations, which became known as the Grenelle 1 and 2 outcomes (Euractiv, 2010), a strategy that applied across 13 sectors. Its labour market plan is the *Green Growth Mobilisation Plan* (Ministère de l'Écologie, 2010). To progress still further, during the environment conference of September 2013 the French authorities committed themselves to integrating sustainable development into all their policies. Since the 1980s, Germany has taken a leading role in climate change mitigation and adaptation. In December 2014, the cabinet adopted the *Action programme on climate protection 2020*. The Programme comprises nine main components, including the 2014 *National energy efficiency action plan*, as well as transport-specific measures, climate-friendly building and housing and a reform of emissions trading. A 2050 version of the programme is expected to be tabled in 2018, and will be updated every three years.

EUROPE

The European Union (EU) describes itself as a family of democratic European countries, committed to working together for peace and prosperity. The organization oversees co-operation among its members in diverse areas including trade, the environment, transport and employment (BBC, 2016).

The evolution of what is now the EU from a regional economic agreement among six neighbouring states in 1951 to today's hybrid intergovernmental and supranational organization of 28 countries across the European continent stands as an unprecedented phenomenon in the annals of history. For such a large number of nation-states to cede some of their sovereignty to an overarching entity is unique. Although the EU is not a federation in the strict sense, it is far more than a free-trade association such as Association of South East Asian Nations (ASEAN), the North African Free Trade Agreement (NAFTA) or Mercosur, and it has certain attributes associated with independent nations: its own flag, currency (for some members) and law-making abilities, as well as diplomatic representation and a common foreign and security policy in its dealings with external partners.

Internally, the 28 EU member states have adopted the framework of a single market with free movement of goods, services and capital. Internationally, the EU aims to bolster Europe's trade position and its political and economic weight. Despite great differences in per capita income among member states (from US$13,000 to US$82,000) and in national attitudes towards issues like inflation, debt, and foreign trade, the EU has achieved a high degree of coordination of monetary and fiscal policies (CIA, 2016).

The 1993 Maastricht Treaty, the treaty that forms the EU, provides for a highly competitive social market economy aiming at full employment, social progress and improvement of the environment (Europa, 2012c). EU Treaties are binding agreements with member states and set out the objectives and rules for the EU institutions and how decisions are made. Under the Treaties, the EU institutions can adopt legislation on which member states are required to act. The EU cannot propose law that is outside the scope of those Treaties. The laws of the EU function alongside the individual laws of each member nation. When there is conflict between the EU member nations' law and EU law, EU law takes precedence (Eupolitix, 2013).

EU legislation on climate change has been characterized by a strategy of cooperation with the international community, compliance with the Kyoto Protocol and a will to maintain leadership through ambitious targets and emission reduction mechanisms. The European Council endorsed the objective of reducing EU emissions of GHGs to 80 to 95 per cent below 1990 levels by 2050. The *Roadmap for moving to a competitive low carbon economy in 2050* (the Roadmap) was adopted by the European Commission in 2011. It describes a cost-effective pathway to reach this objective and gives direction to sectoral policies for all economic sectors, national and regional low-carbon strategies and long-term investments (LSE Grantham, 2015).

The climate change policy that binds every state in the EU is the *2030 climate and energy framework*. The framework sets three key targets for the year 2030, at least 40 per cent cuts in GHG emissions from 1990 levels, at least 27 per cent share for renewable energy and at least 27 per cent improvement in energy efficiency. The main recent legislative and statutory instruments are the 2030 framework for climate and energy policies (strategic document) 2014, European Energy Security Strategy 2014, Fluorinated greenhouse gases (Regulation No. 517/2014 on fluorinated greenhouse gases and repealing Regulation (EC) No 842/2006) 2014, Common Agricultural Policy 2014–2020 2013, Land Use, Land Use Change and Forestry (LULUCF) (Decision No. 529/2013/EU on accounting rules on GHG emissions and removals resulting from activities relating to LULUCF and on information concerning actions relating to those activities) 2013, Energy Efficiency (Directive 2012/27/EU on energy efficiency), amending Directives 2009/125/EC and Emission performance standards for new light commercial vehicles (Regulation (EU) No. 510/2011 setting emission performance standards for new light commercial vehicles as part of the Union's integrated approach to reduce CO_2 emissions from light-duty vehicles 2011 (last amended 2014)), Energy labelling (Directive 2010/30/EU on the indication by labelling and standard product information of the consumption of energy and other resources by energy related products) 2010, Revision of the EU Emission Trading System (EU ETS) (Directive 2009/29/EC amending Directive 2003/87/EC so as to improve and extend the greenhouse gas emission allowance trading scheme of the Community) 2009 (last amended on 30 April 2014), Effort Sharing Decision (Decision No. 406/2009/EC on the effort of Member States to reduce their greenhouse gas emissions to meet the Community's greenhouse gas emission reduction commitments up to 2020) 2009, Geological storage of carbon dioxide (Directive 2009/31/EC on the geological storage of carbon dioxide and amending Council Directive 85/337/EEC, European Parliament and Council Directives 2000/60/EC, 2001/80/EC, 2004/35/EC, 2006/12/EC, 2008/1/EC and Regulation (EC) No 1013/2006) 2009, Emission performance standards for new passenger cars (Regulation (EC) No. 443/2009 setting emission performance standards for new passenger cars as part of the Community's integrated approach to reduce CO_2

Table 3.1 Main elements of the EU INDC submitted to the UNFCCC in 2015

National circumstance	Contribution	Contribution mitigation	Contribution adaptation	Fairness	Review	Means to implement
The EU and its 28 member states	Binding target of at least 40% reduction in GHG emissions by 2030 compared to 1990	Domestic legally binding legislation in place for the 2020 climate and energy package	–	The target represents a significant shift beyond current undertaking of 20% reduction by 2020 and is in line with the EU objective to reduce emissions by 80–95% by 2050 cf 1990	–	–

emissions from light-duty vehicles) 2009 (last amended 2014), Eco-design (Directive 2009/125/EC establishing a framework for the Energy performance of buildings) (Directive 2002/91/EC on the energy performance of buildings) 2002 (Recast adopted by EU Parliament in 2010).

The main elements of the EU INDC submitted to the UNFCCC in 2015 are listed in Table 3.1.

The peak representative organization for business and employers is BusinessEurope and trade unions is the European Trade Union Confederation (ETUC). BusinessEurope has the dual mandate of business association and employers' organization. Both BusinessEurope and the ETUC are active participants with the European Commission and the European Parliament and in the European Economic and Social Committee, which also has responsibility for climate change and environmental policy.

BusinessEurope is a recognized social partner at the European Commission and represents all-sized enterprises in 34 European countries whose national business federations are its direct members. BusinessEurope's objectives are to ensure that the voice of business is heard in European policymaking. To this end it interacts regularly with the European Parliament, Commission and Council as well as other stakeholders in the policy community. It also represents European business in the international arena, ensuring that Europe remains globally competitive.

BusinessEurope supports an ambitious legally binding global agreement, which reflects the long-term objective of limiting global warming below 2°C. It advocates that development of a global carbon market should play a stronger role in the future. Economic instruments can best help to stimulate investment in innovative low-carbon technologies and products in locations where they deliver the greatest. It also fully endorses the EU Emissions Trading Scheme (ETS) as the cornerstone of EU climate policy. It is key to make the ETS work for every sector. This is true for the power sector that requires a carbon price that is meaningful to utilities decisions today and tomorrow. At the same time, the risk of carbon and investment leakage requires keeping strong protection measures for sectors on the global industrial market until main competitors have comparable carbon costs.

Recent published research and reports include: 'On the road to Paris, snapshot of energy efficiency technologies' (BusinessEurope, 2015a), 'Lessons learnt from the current energy and climate framework' (Frontier Economics, 2013), 'A competitive EU climate and energy policy' (BusinessEurope, 2013a), 'A guide on CSR and human rights' (ITCILO, 2015), 'European Commission EU ETS reform proposal' (BusinessEurope, 2016), and 'On the road to Paris. A global deal is our business' (BusinessEurope, 2015b).

European Trade Union Confederation (ETUC) membership is 89 national trade union confederations in 39 countries, plus 10 European trade union federations. The ETUC aims to speak with a single voice on behalf of European workers and have a stronger say in EU decision-making. It also aims to ensure that the EU is not just a single market for goods and services but is also a Social Europe, where improving the wellbeing of workers and their families is an equally important priority. The ETUC believes that this social dimension, incorporating the principles of democracy, social justice and human rights, should be an example to inspire other countries.

The ETUC's action programme 2015–2019 provides that the ETUC will pursue the following objectives:

- The ETUC calls for a change to the European and global economic model, based on long-term investment, a stable but ambitious regulatory framework and a strong social dimension so as to bring about a 'just transition' to a green economy for all Europeans.
- A sustainable investment strategy for Europe.
- No funding for projects at odds with the EU's environmental commitments.

- Development of a low-carbon and sustainable strategy for European industrial policies.
- A just transition policy framework with strong EU financial support, based on the five pillars of social dialogue, investment in quality jobs, greening of education, training and skills, trade union rights and social protection, to tackle climate change (mitigation and adaptation) both at European and international levels.
- An effective European energy community.
- A resource-efficient Europe.
- The greening of the labour market.

Recent published research and reports include: 'ETUC position on the structural reform of the EU emissions trading system' (ETUC, 2015a), 'A new path for Europe: ETUC plan for investment, sustainable growth and quality jobs' (ETUC, 2013a), 'European business, local authorities, civil society and trade unions want EU leaders to live up to their Paris commitments' (ETUC, 2015b), 'ETUC declaration on the Paris Agreement on climate change' (ETUC, 2016), 'ETUC key demands for the Climate COP21' (ETUC, 2015c), and 'Climate change: Implications for employment' (Scott, 2014).

In the global sphere, the ICC and the ITUC are the principal representative organizations of business, employers and workers. Whereas the ICC and ITUC participate as civil society, BusinessEurope and the ETUC have a role as advisors to the European Commission as a Party to the COP. Combined, they are formidable representatives of their constituents who could not otherwise achieve this level of engagement. In EM terms, they serve as strong and effective civil society advocates.

Europe is a case study of this research and is reported in detail at Chapter 5. It is a leader in climate change policy and demonstrates a culture that is sensitive to the environment and all aspects of environmental management, with social and economic development including a mandated role for civil society. In that sense, Europe provides a model of very strong EM.

UNITED KINGDOM

The United Kingdom is made up of England, Wales, Scotland and Northern Ireland. It has a long history as a major player in international affairs and fulfils an important role in the EU, UN and the North Atlantic Treaty Organization (NATO). Britain was the world's first industrialized country. Its economy remains one of the largest, but it has for many years been based on service industries rather than on manufacturing. The process of deindustrialization has left behind lasting social problems and pockets of economic weakness in parts of the country (BBC, 2016).

The UK, a leading trading power and financial centre, is the third largest economy in Europe after Germany and France. Agriculture is intensive, highly mechanized and efficient by European standards, producing about 60 per cent of the country's food needs with less than 2 per cent of the labour force. The UK has large coal, natural gas and oil resources, but its oil and natural gas reserves are declining and the UK has been a net importer of energy since 2005. Services, particularly banking, insurance, and business services, are key drivers of British GDP growth. Manufacturing, meanwhile, has declined in importance but still accounts for about 10 per cent of economic output (CIA, 2016).

This brief introduction to the UK climate change policies and strategies is supported by a major case study at Chapter 6. The UK 2050 Pathway Analysis (DECC, 2010) and the Carbon Plan (DECC, 2011a) are implemented through the Climate Change Act 2008 and the Energy Act 2008. As such, they represent the UK's vision and plans for reducing emissions, climate management and energy efficiency. The long-term perspective offers some certainty for decision-makers against the potential short-termism in policy and facilitates regulatory coherence. These initiatives are considerate of the need for industry to remain competitive and to innovate (Glynn, 2014).

The main legislative and statutory instruments of the government are the Energy Act 2013 (amended 2014), Finance Act 2011: Energy Act 2011 (amended 2014), Feed-in Tariffs for Renewable Electricity 2010 (amended 2012, 2014), Carbon Reduction Commitment Energy Efficiency Scheme 2010 (amended 2013), Climate Change Act 2008, Climate Change and Sustainable Energy Act 2006, Renewables Obligation 2002, Climate Change Agreements 2001 (amended multiple times, most recently 2014), Climate Change Levy 2001 (amended multiple times, most recently 2014), National Adaptation Programme 2013, UK Climate Change Risk Assessment (CCRA) 2012, and the Carbon Plan 2011. The government supported the EU INDC that was submitted to the UNFCCC in 2015 as its intended commitment in the post-Kyoto period. The main elements of the EU INDC are noted in the EU profile above.

The Confederation of British Industry (CBI) is the main representative body for business in the United Kingdom. CBI is a confederation of 140 trade associations, alongside larger and medium-sized businesses who tend to join the CBI directly and represents 190,000 businesses. The CBI's mission is to promote the conditions in which businesses of all sizes and sectors in the UK can compete and prosper for the benefit of all. Function and objectives as set out in the Royal Charter are to provide for

British industry the means for formulating, making known and influencing general policy in regard to industrial, economic, fiscal, commercial, labour, social, legal and technical questions, to act as a natural point of reference to those seeking industry's views and to develop the contribution of British industry to the national economy.

On climate change, the CBI has established the Energy and Climate Change Board with the aim of working with the government to set the right conditions to attract investment in low-carbon solutions and drive consumer demand for sustainable products. At August 2016, the board members include the chief executive of Infinis; chairman of Shell UK; vice chairman of Barclays Corporate; chief executive of BT Consumer; chief executive of Siemens plc; and chief executive of Tata Steel. The Board is supported by professional staff. Recent published research and reports include: 'Setting the bar' (CBI, 2015a), 'Priorities for new government' (CBI, 2015b), 'Effective policy, efficient homes' (CBI, 2015c), 'Small steps, big impact' (CBI, 2015d), 'A climate for growth' (CBI, 2014a), 'Business and public attitudes towards UK energy priorities' (CBI, 2014b), 'Climate change and business: The role of business' (CBI, 2012a), and 'The colour of growth' (CBI, 2012b).

The CBI generates credible and well researched reports, and it communicates regularly with its membership on these issues and declares its opinion is placed before the responsible people in government. That said, the media are not responding and further enquiries within government have failed to establish whether it has traction in the framing of policy.

The UK Trades Union Congress (TUC) is the peak trade union organization, with membership by 52 sectoral and occupational trade unions. The TUC reports that its objective is to raise the quality of working life and promote equality for all. The mission is to be a high-profile organization that campaigns successfully for trade union aims and values, assists trade unions to increase membership and effectiveness, cuts out wasteful rivalry between unions; and promotes trade union solidarity. This broadly reflects the formal objectives recorded at Rule 2 in the Unions Rules and Standing Orders.[1]

The TUC's presence on climate change, sustainability and the environment and in its resources available to the union movement and the public is surprisingly narrow. Its website offers references to its Green Workplaces programme (TUC, 2016a) and the environment (TUC, 2016b), which largely refers back to Green Workplaces, a project initiated in 2009 to train green workplace representatives. It supported the outcome from the UNFCCC Paris COP 21 and called on the government to upgrade its efforts, citing the ITUC commitment to the COP 21 outcome on behalf of the broader union movement.

The TUC', consistent call since the creation of the Green Workplaces programme has been for statutory rights for trade union environment representatives, both in terms of training and facility time. Those rights are to include reasonable time off during working hours to promote sustainable workplace initiatives, carry out environmental risk assessments and audits, consult and receive relevant training. There is nothing to indicate that these demands are achieving traction in industry or with government. Recent published research and reports include: 'Green collar nation: a just transition to a low carbon economy' (Garman and Pearson, 2015); and 'GreenWorkplaces news' (TUC, 2015).

In summary, it is still to be established whether the TUC has moved beyond the first flush of inspiration and it appears government funding was removed in 2009. The website references are based on activities at that time and it is reasonable to expect that any interest and engagement from the membership is a reflection of activities of their employer or local circumstance.

In the context of ecological modernization, the UK has adopted an optimum EM model, subject to constraints; however, effective engagement with all stakeholders across civil society could be enhanced and the process more transparent. It has fully taken up the EU's climate and energy package and has a comprehensive suite of regulations, programmes and accountability. The impact of Brexit on adhering to this optimum model remains to be seen.

FRANCE

France has the Eurozone's second largest economy and is a leading industrial power but has struggled to emerge from the recession since 2008 (BBC, 2016). With more than 84 million tourists per year, France is the most visited country in the world and maintains the third largest income in the world from tourism (CIA, 2016). France today is one of the most modern countries in the world and is a leader among European nations. It plays an influential global role as a permanent member of the United Nations Security Council, NATO, G7, G20, EU and other multilateral organizations. The French economy is diversified across all sectors. The government has partially or fully privatized many large companies, including Air France, France Telecom, Renault and Thales. However, the government maintains a strong presence in some sectors, particularly power, public transport and defence.

Like many of the developed countries in the EU, the French have been positive in their efforts to introduce measures to effectively manage the environmental and climate change issues confronting the continent. It has embraced the EU's climate and energy framework and moved beyond its renewable energy targets. It has also recently announced its intention to include a carbon tax in its autumn 2016 budget, a proposal that would tax EU ETS and non-ETS sectors from 2017 in an effort to set a minimum domestic carbon price of at least 30 euros per tonne. The current traded price for carbon is less than 10 euro (Carbon Pulse, 2016b).

The French government adopted the programmes that emerged from its government-hosted stakeholder consultations, which became known as the Grenelle 1 and 2 outcomes (Euractiv, 2010). The strategy applies across 13 sectors: building, planning, transportation, energy, water, agriculture, biodiversity, health risks, waste, research, consumption, governance and overseas (Euractiv, 2010). Its labour market plan is the *Green Growth Mobilisation Plan* (Ministère de l'Écologie, 2010), which addresses jobs growth and is coordinated across the education, business and regional development portfolios.

The French government's Grenelle Plan-Bâtiment plans the progressive scaling up of capacity to renovate 400,000 units per year by 2013 and 800,000 of the houses that currently use the most energy by 2020. By 2012, all new buildings were to be low power and to be 'positive energy' by 2020 (ADEME, 2011).[2] The draft Bill for the law for energy transition and green growth of 2015 organizes several steering tools to achieve transition, such as the development of renewable energy, development of a recycling economy and energy demand control, especially in the residential and transport sectors (LSE Grantham, 2015).

The French government hosted the UNFCCC 2015 COP (Conference of the Parties), which had the responsibility to agree the next generation treaty to succeed the Kyoto Protocol, a mandate that was fulfilled with the adoption of the Paris Agreement and the Lima Paris Action Agenda.

The principle legislative and policy instruments of the French government are the Farming, Forest and Alimentation Framework No. 2014-1170, Grenelle II 2010, Grenelle I 2009, Energy Policy Framework (POPE, No. 2005-781) 2005, Climate Plan (Policy framework) 2013 and the National Climate Change Adaptation Plan 2011. The government supported the EU INDC that was submitted to the UNFCCC in 2015 as its intended commitment in the post-Kyoto period. The main elements of the EU INDC are noted in the EU profile above.

The peak business association, Mouvement des Entreprises de France (Movement of the Enterprises of France, MEDEF) MEDEF, is also the peak employers' organization. MEDEF represents over 750,000 companies of all sizes throughout the country. MEDEF's 76-member trade federations represent their members with French and European public authorities and in the media to bring attention to the profession's concerns. They analyse and protect their technical, legal and financial interests. They also oversee their sector's collective bargaining agreement and consult with their partners, employee unions, consumer groups and mediation organizations, among others. In 2015 MEDEF was advocating four key reforms: restoring the conditions for the competitiveness of the economy, implementing a real movement to reduce public spending, reconciling business, the economy and French society and setting a course and define a road map for reform.

MEDEF is active on domestic climate change policy and also through BusinessEurope. The three tools for its representations are the Sustainable Development Centre, which aims to make competitiveness central to the energy mix and security of supply policies, promote energy efficiency and the fight against climate change; the Environment Committee, which aims to integrate competitiveness into environmental policies; and the Corporate Social Responsibility Committee. Recently published research and reports include: 'Entreprises et biodiversité comprendre et agir' (MEDEF, 2013), 'Guide sur les initiatives RSE sectorielles première edition' (MEDEF, 2014), 'Joint statement high-level business summit on energy and climate change' (MEDEF, 2015a), and 'Le MEDEF lance le manifeste des entreprises pour la conférence climat Paris 2015 (COP 21)' (MEDEF, 2015b).

The Confédération Générale du Travail (CGT) is acknowledged as the largest of the five trade union confederations in France. It is largest in terms of electoral votes but second largest by member numbers. Membership is open to all employees, women and men, active, private employment and retired, regardless of their social and professional status, nationality, political opinion, philosophical and religion. Its purpose is to defend their rights and professional interests, moral and material, social and economic, individual and collective. Taking into account the fundamental antagonism and conflicts of interests between employees and employers and between needs and profits, the CGT fights capitalist exploitation and all forms of exploitation of wage labour.

On sustainability and climate change it has a dedicated staff appointment, although it still relies on its peak European and international trade unions for leadership and advocacy. On domestic policy it is involved on climate change mainly as it impacts direct workplace and labour management issues.

In the ecological modernization context, France follows an optimum EM model, which is close to excellent, with engagement across all stakeholders and with well-resourced commitments to the regional and international efforts. Its leadership as host of the 2015 COP was a significant contribution to the outcome and is notable for extending engagement beyond the legally binding Agreement, with the other pillars of voluntary commitments by the Parties, finance and NSAs.

GERMANY

Germany is Europe's most industrialized and populous country. Germany has become the continent's economic giant and, as a prime mover of European cooperation Europe's largest economy, is the main player in the EU and a proponent of closer integration. Germany's economic success is to a large extent built on its potent export industries, fiscal discipline and consensus-driven industrial relations and welfare policies. It is particularly famed for its high-quality and high-tech goods (BBC, 2016).

Following the March 2011 Fukushima nuclear disaster, Chancellor Angela Merkel announced in May 2011 that eight of the country's 17 nuclear reactors would be shut down immediately and the remaining plants would close by 2022. Germany plans to replace nuclear power largely with renewable energy, which accounted for 27.8 per cent of gross electricity consumption in 2014, up from 9 per cent in 2000. Before the shutdown of the eight reactors, Germany relied on nuclear power for 23 per cent of its electricity generating capacity and 46 per cent of its base-load electricity production (CIA, 2016).

Germany has taken a leading role in climate change mitigation and adaptation since the 1980s. Traditionally, all political parties support action on climate change and the (non-legally binding) short-term national emission reduction target of at least 40 per cent reduction in GHG emissions by 2020 compared to 1990 levels was reiterated in the 2013 coalition agreement (LSE Grantham, 2015).

In December 2014 the cabinet adopted the *Action programme on climate protection 2020*, which aims to reduce GHG emissions by 62–78 million tonnes CO_2-equivalent by 2020 (as compared to current projections). In the foreword to the Programme, the Minister for the Environment, Nature Conservation, Building and Nuclear Safety, Barbara Hendricks, reported:

> [T]hat cutting greenhouse gas emissions by at least 40 per cent below 1990 levels is the ambitious target Germany has committed itself to for 2020. Our National Inventory Report showed that we had achieved a 24.7 per cent reduction in 2012 and further cuts will follow as a result of measures we put in place before 2014. Dealing with climate change means facilitating and promoting social and economic change in the best possible way. (The German Government's Climate Action Programme 2020 Cabinet decision of 3 December 2014)

The Programme comprises nine main components, including the 2014 *National energy efficiency action plan*, as well as transport-specific measures, climate-friendly building and housing and a reform of emissions trading. A 2050 version of the programme is expected to be tabled in 2018 and will be updated every three years.

While the nuclear disaster in Fukushima, Japan led to the decision to phase out all nuclear power stations by 2022, for a variety of reasons, lignite and hard coal production has been steeply increasing. In order to continue to meet the country's voluntary climate targets, plans for a new law to limit coal-fired generation have been announced. As well, the government last year reached deals with utilities RWE, Vattenfall, and Mibrag to idle eight of their lignite power units and put them in a reserve to support baseload supply shortages, a move which is forecast to cut Germany's greenhouse gas emissions by 11–12.5 million tonnes of CO_2 by 2020 (Carbon Pulse, 2016b).

Adding to its suite of initiatives, the government on 12 May 2016 announced that it will invest €17 billion in a 'broad campaign' to boost energy efficiency over the next five years, with an ultimate goal of halving energy consumption by 2050. The initiative is aimed at helping the country reach emissions reduction targets adopted both domestically and as part of the Paris Agreement.

The plan, dubbed 'Effizienzoffensive', will see the €17 billion spent between 2016 and 2020 and will be comprised of four programmes, which are:

- A competitive tender to find the most cost-effective energy saving measures;
- A pilot programme promoting smart metering;
- An initiative to improve the recovery of waste heat; and
- An initiative to promote cross-cutting technologies, namely those that enhance the efficiency of energy output or its use.

The ministry said a significant increase in energy efficiency is required for the success of the country's the energy transition, or Energiewende, adding that an expansion in renewable energy sources alone won't be enough to hit the country's emissions reduction targets (Carbon Pulse, 2016c).

The main legislative and statutory instruments of government are the Carbon Capture and Storage Act (KSpG) 2012, Grid Expansion Acceleration Act (NABEG) 2011, Energy and Climate Fund Act (EKFG) 2010, Power Grid Expansion Act (EnLAG) 2009, Renewable Energies Heat Act (EEWärmeG) 2008, amended 2009 and 2011, Energy Consumption Labelling Act (EnVKG) 1997, most recently amended 2012, Energy Saving Act (EnEG) 1976, most recently amended 2013, Action Programme on Climate Protection 2020 2014, Energy Concept for an Environmentally Sound, Reliable and Affordable Energy Supply 2010, German Strategy for Adaptation to Climate Change (DAS) 2008, amended 2011. The government supported the EU INDC that was submitted to the UNFCCC in 2015 as its intended commitment in the post Kyoto period. The main elements of the EU INDC are noted in the EU profile above.

In Germany the peak business association and the employer's organization exist separately with discrete mandates. Bundesverband Deutschen Industries (BDI) is the industry organization representing the commercial interests of business and industry in Germany. The BDI is the umbrella organization of German industry and industry-related services. It speaks for 36 trade associations and more than 100,000 enterprises with around eight million employees. Membership is voluntary. Fifteen organizations in the regional states represent the interests of industry at the regional level. BDI has 22 operating divisions and representation from members on BDI committees by over 1,000 people so it is a big and well-resourced organization. The BDI Statutes, at Article 2 and 3 provide that the Association has the task/purpose to protect and foster all common concerns of branches of industries brought together in it. The Association will work with the other leading organizations of entrepreneurship and excluded is the representation of socio-political concerns.

On climate change, BDI has two divisions addressing energy and climate policy, and environment, technology and sustainability, clearly a comprehensive sphere of interest and does not overlook any of the issues. There is no other organization profiled in this research that has committed the level of resources or defined the impact in such all-embracing terms. The many BDI published research and reports include: 'Resource efficiency in the circular economy' (BDI, 2016a), 'Industry supports circular economy initiative' (BDI, 2016b), 'Legal and planning certainty for operators of industrial installations' (BDI, 2016c), 'Improvement proposals for the BREF process' (BDI, 2016d), 'Entrepreneurial freedom vs. further development of medium-related environmental protection' (BDI, 2016e), 'More than 300 different soil types in Europe make one-size-fits-all soil protection legislation difficult' (BDI, 2016f), and 'Bring economy and ecology into line with each other' (BDI, 2016g).

The peak employer's organization is Confederation of German Employers' Associations (BDA), which has as members 14 cross-sectoral regional associations and 51 leading national sectoral federations of employers. The objectives of BDA are to represent the interests of small, medium-sized and large companies from all sectors in all questions linked to social and collective bargaining policy, labour law, labour market, education and societal policy. It actively helps to shape economic framework conditions, pools the voices of employers and lends them weight in the public debate. It has no climate change or environment responsibility.

The peak trade union, the Deutscher Gerwerkschaftsbund Bundesvorstand

(DGB), is the umbrella organization for eight German trade unions. Together, the DGB member unions represent the interests of over six million people. The DGB is the political umbrella organization of the German trade unions and is the voice of working people in Germany. Its objective is to unite and represent the interests of its unions and their members to politicians and other organizations at all levels: from local government to European and international bodies. As a political umbrella organization, the DGB is not involved in collective bargaining, does not organize strikes and does not engage in union activities in workplaces; this work is carried out by its member unions.

There is no indication that DGB has a dedicated climate policy. Rather, it is tailoring its activities on climate change to accord with its other social and labour objectives. In an ecological modernization context, the German experience delivers an optimum model of close to an 'ideal' version of EM with a maximum rating and the overachievement of it ecological objective. While German policy and programmes are based on the EU 2030 climate and energy package, it has committed to domestic programmes that exceed those of the broader European package.

NOTES

1. TUC rules and standing orders. As amended 2010.
2. It has not been possible to establish the performance against these targets.

4. Climate policy in context II: countries outside the EU

Australia, Canada and Singapore are developed countries with internationally competitive advanced economies. However, each is vulnerable to the challenges of climate change. Australia, which is the driest inhabited continent on earth and vulnerable to issues such as floods, drought and bushfire, is frequently criticized for the lack of a national climate change policy and action plan. Canada faces the challenges of conflict between developing its diverse energy resources while maintaining its commitment to the environment. Canada withdrew from the Kyoto Protocol in 2011 and pursued the development of its oil tar sands. The change of government in 2015 promises a change in attitude and leadership on climate change and sustainability policy and to harnessing the authority of its provinces to deliver effective emission reduction programmes. Singapore is a low-lying area that has already experienced rises in sea levels and ambient temperatures. The first Singapore Green Plan was adopted in 1992 and in 2009 it adopted the Sustainable Singapore Green Plan that outlined the sustainable development targets to 2030. The Sustainable Development Blueprint 2015 is the product of feedback obtained from 130,000 people and in 2015 6,000 people were engaged through dialogues, surveys and Internet portals.

In contrast, India and Kenya are developing countries and have maintained a development agenda. India has declared its overriding priorities as economic and social development and poverty eradication. Both are vulnerable to the impacts of climate change and are susceptible to climate-related events. India's climate sensitive sectors are agriculture, water and forestry, while Kenya is experiencing irregular and unpredictable rainfall resulting in drought during the long rainy season and other regions are experiencing floods during the short rains. The Indian government adopted a climate change action plan in 2008 that was revised in 2014. Its 12 missions are solar, enhanced energy efficiency, sustainable habitat, water, sustaining the Himalaya ecosystem, green India, sustainable agriculture, strategic knowledge for climate change, wind energy, health, coastal resource and waste to energy. Earlier this year, 2016 the World Bank Group signed a joint declaration with India to collaborate further on increasing solar energy use and will provide more than US$1 billion to help India's solar plans. Kenya's Constitution requires the attainment of ecologically sustainable development. In 2010, the Ministry of Environment and Natural Resources (MENR) launched the National Climate Change Response Strategy and later the 2013–2017 Climate Change Action Plan. Over the five years to 2015 the MENR has produced several drafts of the National Environment Policy that recognizes climate change as one of the direct causes of natural disasters and proposes measures to address climate change.

AUSTRALIA

Australia ranks as one of the best places to live in the world by all indices of income, human development, health care and civil rights. The sixth largest country by landmass, its comparatively small population of approximately 23 million people in 2016 is concentrated in the highly urbanized east coast of the continent. The island combines a wide variety of landscapes, deserts in the interior, hills and mountains, tropical rainforests and densely populated coastal strips with long beaches and coral reefs off the shoreline (BBC, 2016).

Australia has become an internationally competitive, advanced market economy and its location in one of the fastest growing regions in the world. Long-term concerns include an aging population, pressure on infrastructure and environmental issues such as floods, droughts, and bushfires. The services sector is the largest part of the Australian economy, accounting for about 70 per cent of GDP and 75 per cent of jobs. Australia plays an active role in the World Trade Organization, APEC, the G20 and other trade forums (CIA, 2016).

Australian legislation and programmes to address this vulnerability to climate change are wide ranging, as are detailed below. Professor Garnaut (2011), in his report to the Australian government, observed that Australian climate policy was being reflected in a range of ad hoc programmes at a cost of approximately AU$1 billion per annum. The analysts Energetics, however, have reported that the current suite of climate policies, including a not yet finalized National Energy Productivity Plan, can cut GHG emissions by a total 960 million tonnes of CO_2 by 2030, enough to meet the nation's 26 to 28 per cent below 2005 reduction target (Carbon Pulse, 2016a).

Climate change has been a contentious issue in Australia, with controversy over the introduction of federal legislation to limit GHG emissions becoming particularly acute in 2009 with the two major political parties advocating different approaches. Domestic policy for and against pricing carbon and an emission trading system (ETS) has been argued on political party lines, the differences exaggerated by intra party turmoil and political instability since 2011, with the country changing prime minister five times in five years, which included two elections that resulted in changes to the government. Since ratifying the Kyoto Protocol in 2007, the government has been engaged internationally and in 2012 with the European Commission announced their intension to connect the Australian and EU ETS. The change of government in late 2013 stalled the implementation of the new legislation and was subsequently repealed (LSE Grantham, 2015).

In the wake of the 2015 Paris Agreement on Climate Change, Australian activists and other stakeholders have been expressing their views publicly. Melbourne University's Professor Christoff submitted a six-point plan for the candidates of the 2016 federal election, advocating that climate change should be an issue with bipartisan support[1], as is the case in many developed countries, rather than the subject of the present ideological wedge politics, phobia that has been attached to the prospect of carbon pricing should be overcome, and discussion should be about the positives of action rather than political and community pain. He also recommends a climate target 'with teeth', noting that Australia's current target is among the weakest of all developed countries alongside Canada and New Zealand and points out that with the current implemented policy measures Australia's emissions are set to increase to more than 27 per cent above 2005 levels (the INDC committed to reduce emission by 26 to 28 per cent below 2005 levels by 2030) (Christoff, 2016). These concerns were also expressed separately by the academic Tim Stephens, who observed that whilst the government's active and supportive participation in the Paris climate negotiations signaled that, on the international plane at least, some bipartisanship has returned to Australia's climate policy, on the domestic front the major parties are still at loggerheads (Stephens, 2016). At the ceremony to sign the Paris Agreement held in New York on 22 April 2016, the then Australian Minister for the Environment Greg Hunt announced the government's plans to ratify the Agreement (IISD, 2016).

The main statutory and policy instruments are the Carbon Farming Initiative Amendment Bill 2014, Greenhouse and Energy Minimum Standards Act 2012, Clean Energy Finance Corporation Act 2012 Australian National Registry of Emissions Units Act 2011, Climate Change Authority Act 2011, National Greenhouse and Energy Reporting Act 2007, and the National Strategy on Energy Efficiency 2010. The main elements of the INDC submitted to the UNFCCC in 2015 are shown in Table 4.1.

The Australian Chamber of Commerce and Industry (ACCI)

Table 4.1 Main elements of the INDC submitted to the UNFCCC in 2015

	National circumstance	Contribution	Contribution mitigation	Contribution adaptation	Fairness	Review	Means to implement
Australia	–	Reduce GHG emissions by 26–28% below 2005 by 2030	The Emission Reduction Fund supports businesses to reduce emissions, supported by the Renewable Energy Target	To develop a National Resilience and Adaptation Strategy	The target doubles the rate of emission reduction, is a significant increase beyond the 2020 target, and is comparable with other advanced economies	–	Market mechanisms through the Emission Reduction Fund

membership is eight state and territory Chambers of Commerce and 49 national industry associations. There are no direct corporate members. ACCI's core service is private sector advocacy, representation and policy development on national and international matters that impact business. It embraces business advocacy, from trade and commerce, economics and tax, to employment, labour and social policy. Climate change issues are addressed through the Sustainability Committee, which has the mandate to consider national and international issues pertaining to sustainability, including climate change policy and regulation. The committee is to identify issues of importance for business and industry, to be an avenue for information exchange and consultation and to provide advice and feedback to help develop solutions on relevant policy issues relating to business. It is also to inform the ACCI policy agenda as being representative of the business community. It is required to participate in the ACCI political engagement and advocacy activities. A priority is a watching brief on the UN climate change negotiations and the impacts this could have on Australian targets and policy measures including a scheduled review of the Emission Reduction fund. The committee is interested in ways to facilitate the government's Clean Energy Finance Corporation to finance smaller energy efficiency projects.

The Sustainability Committee has initiated responses to a number of government enquiries, including the Australian 2030 emissions targets that formed the basis of the government's INDC, and one-stop shops for environmental approvals in each state. It has made submissions to the other political parties about climate policies.

The Australian Council of Trade Unions (ACTU) is the peak union body representing almost two million members in 46 unions. Its stated objective is the socialization of industry. There is no apparent dedicated domestic climate change activity. The ACTU does not have a publicly stated policy position on climate change, beyond its support for the principle of a just transition and a decent work outcome for those affected, the position of its international affiliate, the ITUC. While respected, informed and professional advocates on the labour relations and labour market issues impacting their members, the ACTU relies on its international affiliates to advocate their climate change interests and it is only when domestic policy requires their intervention will they act.

Both the ACCI and ACTU are constrained by limited resources which must be applied in a strategic manner and according to priorities which are often influenced significantly by the immediacy of the impact on their membership.

In the context of ecological modernization, overlaying the EM template on Australia's INDC rates only 25/50, which is likely to yield a sub-optimum outcome, as all elements of EM are actioned only in a modest fashion. The government's commitment under the INDC submitted to the 2015 Paris COP is considered at the lower end of the scale for developed countries and the actions across the state, industry, civil society and ecological consciousness are modest relative to other countries. The Australian community and industry are well off and insulated from the direct impacts of the early onset of climate change with an abundance of cheap fossil fuels and solar energy, infrastructure to store water in the instance of drought and a large unusable land mass so that issues surrounding waste disposal and air pollution are absent. The major concern to consumers is for bayside landowners whose properties are threatened by coastal erosion.

CANADA

Canada is the world's second largest country by surface but is relatively small in terms of population. It is one of world's top trading nations and one of its richest. Alongside a dominant service sector, Canada also has vast oil reserves and is a major exporter of energy, food and minerals (BBC, 2016).

Canada faces the challenges of meeting public demands for quality improvements in health care, education, social services and economic competitiveness as well as responding to the particular concerns of predominantly francophone Quebec. It also aims to develop its diverse energy resources while maintaining its commitment to the environment. Alberta's oil sands development has significantly boosted Canada's proven oil reserves. In 2016 Canada ranked fifth in the world of proved oil reserves and is the largest foreign supplier to the US of energy, including oil, gas, and electric power, and a top source of uranium imports (CIA, 2016).

Canada has no comprehensive federal climate legislation. An act to implement Canada's targets under the Kyoto Protocol during the first commitment period of 2008–2012 was introduced in 2007. However, in 2011, Canada announced it would withdraw from the Kyoto Protocol and officially repealed the Act in 2012.

The Conservative government's climate change plan 'Turning the corner: action plan to reduce greenhouse gases and air pollution' was announced in 2007 and provided the groundwork for Canada's approach to tackling climate. It expressed the priority of realigning policies and regulations in order to maintain economic prosperity while protecting the environment and harmonizing the regulatory framework with the US, its largest trading partner.

The principal policy positions of the government were the Action on

Climate Change and Air Pollution (2007) and the Federal Adaptation Policy Framework (2011). These policies remain the framework for action by the government. The October 2015 federal election delivered a change in government from Conservative to the Liberal Party of Canada under the Prime Ministership of Justin Trudeau and with that a more liberal policy approach to social and environmental issues. At this time, though, little has been introduced in the form of solid and published policy and legislation to parliament.

The legislative instruments of government are limited and include the Canada Foundation for Sustainable Development Technology Act (S.C. 2001, c. 23), Canadian Environmental Protection Act 1999 (CEPA, 1999) (S.C. 1999, c. 33) 2000, and the Energy Efficiency Act (S.C. 1992, c. 36) last amended in 2008. Despite the lack of comprehensive federal legislation, provinces have been active in passing their own climate legislation. At the Lima COP 20 in 2014 the provinces of Ontario, Quebec and British Columbia issued a joint statement with California to lead international actions to fight climate change and collaborate for an international agreement at the Paris COP 21.

The target of the Action on Climate Change and Air Pollution (Environment Canada, 2007) was a total reduction of Canada's GHG emissions by 20 per cent by 2020. They claimed this was one of the most stringent targets in the world. The policy requires all major industry sectors to respect the government's aggressive limits to reduce GHGs, that the government will act to reduce emissions from cars and trucks, will increase the range of energy-efficient products and will improve air quality. The policy also commits the government to work with provincial and territory governments, NGOs, communities and individual Canadians. Another policy measure is the creation of a climate change technology fund financed by levy on emissions and the trading of carbon credits on domestic and international markets; however, at the time of writing this book this has not yet been implemented.

The Federal Adaptation Policy Framework was adopted in 2011 and remains current, although there is little evidence that it has been aggressively implemented. The Framework describes the context as being that the impacts of climate change are evident in every region across Canada. Higher temperatures, declining sea and lake ice, diminishing glaciers, melting permafrost, more heat waves, more violent storms and increased coastal erosion are being observed. The north is particularly vulnerable and is experiencing changes that are more extreme and occurring faster than the rest of Canada.

The policy recognizes that the climate change impacts are not only physical, they can have long lasting economic, social, environmental and human health effects. It notes the benefits, that Canada will also experience longer agricultural and ice-free shipping seasons and expanded tourism and recreation opportunities (Environment Canada, 2011).

The new liberal government has announced the proposed revised policy template under the banner of *Canada's way forward on climate change* (Environment Canada, 2016). The government will:

1. Contribute to global efforts by working with global partners on the ambitious global agreement, and supporting the poorest and most vulnerable countries.
2. Collaborate with Provinces and Territories within 90 days of the Paris outcome to establish a Pan-Canadian framework for combatting climate change and set a national target while ensuring they have the flexibility to design their own carbon pricing policies.
3. Invest in Clean Energy and Clean Technology through a CA$2 billion low carbon economy trust, fulfill the G20 commitment to phase out fossil fuel subsidies and protect energy security (Environment Canada, 2016).

The main elements of the government's INDC submitted to the UNFCCC in 2015 are shown in Table 4.2.

Business associations, regional and provincial employers' organizations[1] and trade unions are informed and committed activists and participants in the domestic and international climate change debates, more so than many of the peak organizations in other countries. However, absent from their policy platforms are the employment and labour market impacts beyond the broader trade union mantra of a just transition and decent work for affected workers. Important to note here is that the business associations, employers' organizations and trade unions in the provinces and regions hold great authority in that they often act separately from the peak national organizations and make representations to government. The vast distances and areas impose regionally unique concerns that are often difficult to address within a national context and policy. The example to best illustrate this is the representation of Canada at the International Organisation of Employers. The representative is a consultant engaged to monitor activities and, while a respected participant, he has no mandate to advocate on behalf of Canadian employers, for the reason that a peak employers' organization in Canada does not exist.

The main representative organization for business in Canada is the Canada Chamber of Commerce. Membership is 450 chambers of commerce and boards of trade, representing 200,000 businesses in all sectors. Its objectives are to advocate for public policies that will foster a strong, competitive economic environment that benefits businesses, communities

Table 4.2 Main elements of the INDC submitted to the UNFCCC in 2015

	National circumstance	Contribution	Contribution mitigation	Contribution adaptation	Fairness	Review	Means to implement
Canada	Growing population, extreme temperatures, large landmass and significant natural resources	Reduce GHG emissions by 30% below 2005 levels by 2030	Has established stringent coal-fired electricity standards and stringent GHG emission standards for the transport sector	–	–	–	The government has in place legislative instruments coupled with significant investments in clean energy technologies

and families. On climate change, the Chamber says it supports evidence-based policymaking that appropriately accounts for environmental externalities as well as efforts by the government of Canada to cooperate with provinces and territories to address environmental issues that are of shared jurisdiction. It favours a price on carbon, supports the creation of a water strategy and believes in the imperative to foster technological innovation and ensure efficient regulatory processes. Recently published reports and research include: 'What does COP21 mean for Canadian business?' (CCC, 2015), 'The measures that matter: how Canada's natural resource sector is working to protect the environment' (CCC, 2014), and '$50 million a day' (CCC, 2013).

Canadian Labour Congress (CLC) membership is 90 national and international unions, provincial and territorial federations of labour and community-based labour councils that represent 3.3 million workers. Its objectives are to advocate politically for policies and programmes that improve the lives of all Canadians, such as the creation of better and more secure jobs, better public pension plans and retirement security, a stronger public health care system, affordable and accessible child care and to advocate in parliament and in the courts to advance legislation that improves the day-to-day lives of all Canadians, such as workplace safety and collective bargaining rights and employment equity.

On climate change, the CLC comments that Canadian unions are committed to playing their part in the fight to slow global warming and keep the planet viable for future generations. The priority is working with employers and governments to ensure a just transition to a carbon-free economy that supports displaced workers and creates millions of good, green jobs. Recently published research and reports include: 'David Suzuki joins with CLC to support a One Million Climate Jobs plan' (CLC, 2016a), 'One million climate jobs a challenge for Canada' (CLC, 2016b), 'Making the shift to a green economy' (CLC, 2015d), 'Collaborative approach will be key to realizing Canada's climate change obligations' (CLC, 2015a), 'Labour delegation to champion a just transition to a green economy at Paris Climate Change Summit' (CLC, 2015c), 'CLC report: reducing greenhouse gas emissions in Canada' (CLC, 2015e), and 'ENERGY – alternatives for a green economy' (CLC, 2015b).

In an ecological modernization context, Canada would appear to have an optimum EM model, with effective contributions to the process by sections in the model. The recent change of government promises to improve the country's effectiveness on climate change mitigation and adaptation, and leadership; however, at present it is rated on the basis of its varied commitment to the international process having withdrawn from the Kyoto Protocol in 2011, the exploitation of its coal tars sands find without an effective mechanism in place to contain the GHG emissions, noting that it is investing heavily in the development of carbon capture and storage technology.

SINGAPORE

Singapore is a wealthy city state in south east Asia. Once a British colonial trading post, today it is a thriving global financial hub and described as one of Asia's economic 'tigers'. It is also known for its conservatism and strict local laws and the country prides itself on its stability and security. Densely populated, most of its people live in public housing tower blocks. Its trade-driven economy is heavily supported by foreign workers. In 2013, the government forecast that by 2030, immigrants will make up more than 50 per cent of the population (BBC, 2016).

Singapore has become one of the world's most prosperous countries with strong international trading links (its port is one of the world's busiest in terms of tonnage). Singapore has a highly developed and successful free market economy. It enjoys a remarkably open and corruption-free environment, stable prices and a per capita GDP higher than most developed countries. Unemployment is very low. The economy depends heavily on exports, particularly consumer electronics, information technology products, medical and optical devices, pharmaceuticals, and on its vibrant transportation, business, and financial services sectors (CIA, 2016).

Singapore is a low-lying area that has already experienced rises in sea levels and ambient temperature. The possible effects of changing weather patterns in Singapore include accelerated coastal erosion and higher incidence of intense rain or prolonged drought. Climate change will also affect biodiversity of plants and animals, and their greenery. Singapore may also experience disruptions to food supplies, and business supply chains if trading partners are affected by extreme weather events (Republic of Singapore, 2012).

Singapore has always placed a high priority on environmental issues as part of its aim to create a clean and green garden city for its people. After gaining independence from the British in 1965 Singapore has pursued the concurrent goals of growing the economy and protecting the environment. Singapore ratified the UNFCCC 1997 Kyoto Protocol as a non-Annex 1 country. In 2007 an Inter-ministerial Committee on Climate Change (IMCCC) was set up to oversee inter-agency coordination on climate change. In 2010 the National Climate Change Secretariat was established as a dedicated unit under the prime minister's office to develop and provide coordination for climate change-related policies. It published the National

Climate Change Strategy in 2012 and created agencies to effectively implement the strategy. In 2015 it adopted the Sustainable Singapore Blueprint after consultation with 6,000 people from the community.

Singapore is active in the region. It is a member of Association of South East Asian Nations (ASEAN). It is encouraging efforts to develop an ASEAN climate change initiative and to develop regional strategies to enhance capacity for adaptation, a low carbon economy and promote public awareness to address effects of climate change (LSE Grantham, 2015).

The first Singapore Green Plan was adopted in 1992 and in 2009 the Sustainable Singapore Green Plan, which outlined the sustainable development targets to 2030, was adopted. The government adopted the National Climate Change Strategy (NCCS) in 2012. The core of the Strategy is greater public transport usage, improved energy efficiency in buildings, deeper behavioural adjustments and changes to business processes, and research and development. The guiding principles are long-term planning, pragmatic and economically sound measures, the pursuit of the economic and environmental objectives together, harnessing market forces, developing innovative solutions and institutional reform through the National Climate Change Secretariat. Singapore is also developing its solar and photovoltaic technology capability, which will help facilitate the deployment of solar power on a large scale.

The Strategy acknowledges the requirement to develop the human capital necessary for the proposed research and development programme, a limited but nevertheless important intervention in the nation's labour market plan. The government's early action on climate change and the environment and its small land mass has left little capacity to reduce emissions further or address its sustainability targets in the absence of advances in technology.

A complement to the NCCS is the Sustainable Singapore Blueprint (MEWR, 2015). The principles of the Blueprint that build on the work of the original blueprint are eco-smart endearing towns, a car-lite Singapore, moving towards a zero-waste nation, a leading economy and an active and gracious community. This edition of the Blueprint is the product of feedback obtained from more than 130,000 people through recent initiatives, including the Land Transport Master Plan 2013 and the Urban Redevelopment Authorities Master Plan 2014. Civil society engagement and ecological consciousness in Singapore is an exemplar, essential for the achievement of a strong ecological outcome and sound policy (MEWR, 2015).

While Singapore is not now or in the past contributing materially to the GHG emissions problem, it is dependent on other countries for energy, water and internally it does not have a capacity for waste management or the space for a vibrant private transport sector and the consequent issues of fuel, pollution and roads versus gardens. It has taken a positive approach to finding long-term solutions to these issues and intends to become a supplier to the rest of the world of the technologies developed to address its domestic requirements.

Singapore presents a unique situation in terms of development, culture and social attitude. This is reflected in the work of the representative organizations for business and workers but there is no integration of the impact of climate change on the workplace and labour market preparedness is absent.

The main legislative and regulatory instruments of government in respect of climate change are the Energy Conservation Act (Chapter 92C) 2012, last revised 2014, National Environment Agency Act (Chapter 195) 2002, last revised 2003, Energy Market Authority of Singapore Act (Chapter 92B) 2001, last revised 2012, Electricity Act (Chapter 89A) 2001, last revised 2006, Gas Act (Chapter 116A) 2001, last revised 2008, Building Control Act (Chapter 29) 1989, last revised 2012, Sustainable Singapore Blueprint 2015, National Climate Change Strategy June 2012. The main elements of the governments INDC submitted to the UNFCCC in 2015 are outlined in Table 4.3.

Like Germany, the peak employers' and business organizations coexist with clearly defined and non-overlapping mandates.

The employers' organization, the Singapore National Employers Federation (SNEF), has a membership of 3,000 businesses. Its mission is to advance tripartism and enhance labour market flexibility to enable employers to implement responsible employment practices. The SNEF focus areas and services include representing the key interests of employers in national tripartite committees, forums and national-level reviews; providing consultancy and advice to corporate members on the proper application of local labour laws, policies and tripartite guidelines; keeping members informed on developments in labour, manpower and employment issues through briefings, industrial group meetings and other platforms; facilitating employers' efforts to build an inclusive workforce and progressive workplaces; and providing timely research and information. It has no apparent climate interest.

The Singapore Business Federation (SBF) has a membership of 22,500 companies as well as key local and foreign business chambers. Under the government's Singapore Business Federation Act, all Singapore-registered companies with share capital of S$0.5 million and above are members of SBF. Its objective is to serve as the bridge between Singapore's business community and the government and it is the key provider of capability-building initiatives and services for Singapore businesses. Locally, SBF

Table 4.3 Main elements of the INDC submitted to the UNFCCC in 2015

	National circumstance	Contribution	Contribution mitigation	Contribution adaptation	Fairness	Review	Means to implement
Singapore	Urban density, limited land mass, low flat land, low wind speeds, lack of geothermal resources	Reduce emission intensity (per S$ of GDP) by 36% from 2005 levels by 2030 and stabilize emissions aiming to peak by 2030	Limited options given the natural circumstance, very energy efficient	Programmes in place to address food security, infrastructure resilience, public health, flood risks, water security and coastline	Ambitious target given the limited options for action	–	Domestic sources

promotes engagement between Singapore businesses and the government through various initiatives, including the SME Committee (SMEC) – a platform for small and medium enterprises (SMEs) to raise their concerns with policymakers – and regular business climate surveys. Internationally, SBF also champions its members' needs on the global stage through participation in prominent international business forums. These include the ASEAN Business Advisory Council, Asia Pacific Economic Cooperation Business Advisory Council, B20 and the International Chamber of Commerce (ICC). It also has no declared interest in climate change.

Both business organizations, despite no demonstrable interest in climate change, have declared a commitment to sustainable development and CSR. They refer to triple bottom line reporting and corporate social responsibility but they do not talk climate change or activism in policy or broader civil society organization engagement. The SBF was established by an Act of Parliament and membership is compulsory, to ensure proper governance and controlled behaviour in both domestic and international markets. Membership of the SNEF is voluntary and therefore there is more latitude in their response the requirements of members. The SNEF is a founder of the Sustainable Development Business Group but otherwise its mandate is human resources and industrial relations.

The Singapore National Trades Union Congress (SNTUC) is a confederation of 60 affiliated unions, one taxi association, 11 social enterprises and six related organizations. The objectives of the SNTUC are quite different to most unions: to serve as a centre for democratic non-communist trade unionism, to help Singapore stay competitive and workers remain employable for life, to enhance the social status and wellbeing of workers and to build a strong, responsible and caring labour movement. The modern trade union movement in Singapore was formed out of the conflict that led to the alignment with Malaysia against the forces of the communist countries and the communist unions in Singapore in the 1960s. That led to a modernization programme that saw the forming of cooperatives in the 1970s that have resulted in the present SNTUC as responsible for the largest supermarket chain in the country ('FairPrice Co-operative Ltd, the largest grocery retailer in Singapore. Our E-commerce portal provides a wide range of groceries from fresh produce to electronics'), with holiday resorts and an Institute of Labour Studies. SNTUC programmes for members include career activation, career coaching, inclusive growth, legal clinics, migrant workers' centre help kiosks, migrant workers' centre student outreach, progressive wage model, shared parenting programme, women's network, workplace health and safety programmes, and union training (NTUC, 2016).

The concept of tripartism has served the country well with prosperity being shared, wages growth and wage restraint are equally adopted as campaigns of the Union, depending on the economic circumstances. SNTUC's media profile is strongly oriented to its member benefit programmes through it Fairprice supermarkets and campaigns like plastic bag usage, cigarette labelling and then through its care packages.

Although, the business associations, employers' organizations and trade unions do not declare an interest in climate change; the parallel economic and environmental goals of the country since independence must also be reflected in the culture of the organizations. As well, the pervasive nature of the government's strategies to address the country's necessary transition to sustainability must also impact their constituents.

The Singapore Branch of the Global Compact,[2] Singapore Compact was co-founded by SNEF and SNTUC, and is the national society that promotes sustainable development for businesses and stakeholders and continues to expand its reach and programme line up. In recent years, discussions around CSR and sustainability have also been gaining considerable traction locally. In 2014, the Singapore Stock Exchange announced its plan to adopt the 'comply or explain' approach for sustainability reporting for all listed companies.

At March 2015 the Compact membership stood at 933, which included about 483 corporate, institutional and associate members and 406 members from the youth wing. Members consist of a diverse range of companies including multinational corporations (MNCs), SMEs, as well as trade unions, CSR consultants and academics. The Singapore Compact is a society of like-minded entities coming together to share with and learn from one another best practices that it is expected will create a more sustainable and responsible corporate environment in Singapore (Singapore Compact, 2014).

The Singapore model represents an optimum EM model, given the country's constraints. Its INDC declares a strong commitment to contributing to the global effort and achieving ambitious targets. It has committed the resources necessary to achieving those targets with a strong regulatory platform, a technology-based approach and with the engagement of civil society and the community.

INDIA

India is the world's largest democracy and the second most populous country, with many languages, cultures and religions, which makes it highly diverse. It is still tackling huge social, economic and environmental problems, corruption and poverty are widespread and the vast mass of rural population remains impoverished (BBC, 2016). Despite these problems, economic growth following the launch of economic reforms in

1991 and a massive youthful population are driving India's emergence as a regional and global power (CIA, 2016).

The government's National Action Plan on Climate Change (2008) defines the challenge as sustaining rapid economic growth while dealing with the global threat of climate change. The economy is closely tied to its natural resource base and climate-sensitive sectors such as agriculture, water and forestry, which may be adversely impacted by climate change. However, its overriding priority remains economic and social development and poverty eradication, a point the present prime minister (at 2016) reinforced during the negotiations for the 2015 Paris Agreement. India's climate risk assessment in the second communication to the UNFCCC states that climate change, leading to recession of glaciers, decrease in rainfall and increased flooding could threaten food and water security, put at risk natural ecosystems and species that sustain the livelihood of rural households and adversely impact coastal systems due to sea level rise and increased extreme events.

India was a non-Annex 1 country under the Kyoto Protocol and thus had no binding target for emission reduction. It is an active participant in the Clean Development Mechanism (CDM) established under the Protocol. It has also created the National Clean Energy Fund, funded by a levy on coal to finance and promote clean energy (LSE Grantham, 2015).

The National Action Plan on Climate Change (GOI, 2008) was adopted by the government of India in 2008 and remains the national policy. The Plan is a model of strong ecological modernization. It notes in its preamble that development hinges on new technologies, appropriate institutional mechanisms, public–private partnerships and civil society action, and promoting understanding of climate change, adaptation and mitigation. The eight missions of the Plan are solar, enhanced energy efficiency, sustainable habitat, water, sustaining the Himalaya ecosystem, green India, sustainable agriculture and strategic knowledge for climate change (GOI, 2008). Four new missions were announced in 2014 and are pending approval: wind energy, health, coastal resource and waste to energy (LSE Grantham, 2015). The Indian government is advised on these missions by an advisory council chaired by the prime minister and has broad based representation from key stakeholders including government, industry and civil society.

In the INDC submitted to the UNFCCC as the commitment to the Paris Agreement (UNFCCC, 2015), the government described its situation in the terms that India supports 17.5 per cent of the world population and in that it has the largest proportion of global poor with 363 million people living in poverty, 92 million without access to safe drinking water, 304 million without electricity and 300 million relying on solid biomass for cooking. To that end, its priority is economic development and poverty eradication. Further, its commitment to reduce the emissions intensity of GDP by 33–35 per cent by 2030 from 2005 level was contingent on an ambitious global agreement and additional means provided by developed countries. The finance required for adaptation plans was US$206 billion between 2015 and 2030 with further assistance required for strengthening resilience and disaster management estimated at $7.7 billion to the 2030s, and $834 billion for mitigation till 2030 (GOI, 2015).

In the week leading up to the Paris COP 21, the Indian prime minister was an active advocate and campaigner for India. He dampened expectations when at the G20 Summit in the month before the COP he blocked efforts to a strong statement by the G20 members towards an ambitious climate accord, saying India did not want the G20 to interfere in the Paris negotiations (Barker and Clark, 2015). On the opening day of the COP he went further, demanding that poor nations had the right to burn carbon to expand their economies, that climate change was not of their making and insisting that emission reductions in his country must come paired with billions of dollars of investment by the developed world (Davenport and Harris, 2015). This message was reinforced by the BRIC countries of Brazil, Russia, India, China and South Africa who called a news conference in the middle of the second week of the COP to convey the message that they are big but not rich and it is the rich countries that should pay to address global warming (Chan, 2015).

During the COP, the Indian and French governments unveiled a plan to mobilize more than US$1 trillion to make solar energy affordable in sun-rich developing countries through the International Solar Alliance, which aimed to sign up 121 countries including the US and China as well as a long list of developing countries (Clark and Stothard, 2015).

Clearly some nations' positions are driven by their own self-interest. For India, it has a need to increase its electricity generation using coal and renewables and it needs to finance this development. Prime Minister Narendra Modi used the forum of the COP to launch an ambitious programme to accelerate investment in renewables (Krauss and Bradsher, 2015). However, while the principle was applauded, the Modi model was opposed by the US government and subsequently by the WTO which ruled against India, saying the power purchase agreements with solar firms were inconsistent with international norms and were discriminatory (Press Trust of India, 2016). The viability of the solar programme is further challenged on the grounds of commercial viability, a factor of the market where competition has delivered purchase prices at levels close to the cost of coal-generated power and less than the cost of production, making investors reluctant and expectant that contactors will not be able to deliver (Mallet, 2016).

While the government has a vision and a realistic plan, it also is confronted by many challenges in that the plan is dependent on international support in terms of finance and technology. Its plans are necessarily ambitious but optimization of the domestic benefits could bring the government and delivery expectation into conflict with the rules of the global marketplace from which it will expect the financial and technical resources. And, while the government's policy platform and the Advisory Council embrace engagement with stakeholders including government, industry and civil society, business and employers' organizations and trade unions are largely absent.

The legislative and statutory instruments of government are the Finance Bill 2010–11 and the Clean Energy Cess Rules, 2010, Electricity Act 2003, amended in 2007, Energy Conservation Act 2001, amended 2010, National Electricity Plan (Generation) 2012, National Policy on Biofuels 2009, National Action Plan on Climate Change 2008, Energy Conservation Building Code 2007, Tariff Policy 2006 amended 2011, Integrated Energy Policy 2006, National Electricity Policy 2005. The main elements of the governments INDC submitted to the UNFCCC in 2015 are outlined in Table 4.4.

Employers' organizations and trade unions in India are relatively well-resourced as CSOs and as representatives of sections of society. The research for this book selected the largest but not the only peak organizations. The magnitude of the population and land mass and its developing economy situation mitigates against the centralization of representation and accordingly the capacity of the organizations for strategic and effective intervention with the central government.

India has a proliferation of peak business representative organizations embracing the sectoral and regional groups. The International Organisation of Employers (IOE) has four Indian employers' organizations as members, the All India Organisation of Employers (AIOE), Employers' Federation of India (EFI), Standing Conference of Public Enterprises (SCOPE) and the Council of Indian Employers (CIE). The Federation of Indian Chambers of Commerce and Industry (FICCI) provides the secretariat for the CIE. The CIE is the formal representative organization at the IOE for the AIOE, EFI and SCOPE. None of the organizations has any stated policy on climate change or any apparent interest in the impacts of climate change policy on the labour market.

The two largest of the business representative organizations are the Federation of Indian Chambers of Commerce and Industry (FICCI) and the Confederation of Indian Industry (CII).

The FICCI membership is 250,000 businesses and has as its objectives to be the voice of India's business and industry, from influencing policy to

Table 4.4 Main elements of the INDC submitted to the UNFCCC in 2015

	National circumstance	Contribution	Contribution mitigation	Contribution adaptation	Fairness	Review	Means to implement
India	17.5% of world's population, 300 million live in poverty, 300 million without electricity and safe drinking water	Reduce emission intensity of GDP by 33% by 2030 from 2005 level	40% of electric power from non-fossil fuel sources by 2030 Create carbon sink of 2.5–3 million tonnes by 2030	Resilient agriculture Conservation of water Coastal zone management Disaster management	Considered fair and ambitious given developmental challenges. Conditional on international assistance	–	Ambitious global agreement International assistance requirements for adaptation US$206 billion to 2030, $7.7 billion for disaster management, $834 billion for mitigation to 2030

encouraging debate and engaging with policymakers and civil society to articulate the views and concerns of industry. Internationally it is an active participant in the International Chamber of Commerce (ICC), and on employment matters in the International Organisation of Employers (IOE) through the All India Organisation of Employers (AIOE). On climate change, FICCI believes that it is essential to secure the proactive participation and involvement of businesses and people for improving environment quality. It observes that the adoption of clean, climate and health-friendly technologies in every sphere of activity is of paramount importance in enabling environmental improvement. FICCI's Environment and Climate Change Division has taken up a broad spectrum of initiatives to address industry's issues pertaining to environment and climate change, and also works on environmental projects with various national and international agencies, organizes outreach events, training programmes and workshops for industry awareness on issues pertaining to environment and climate change.

The Confederation of Indian Industry (CII) membership is 8,000 businesses from the private as well as public sectors, including SMEs and MNCs, and an indirect membership of over 200,000 enterprises. Its objective is to create and sustain an environment conducive to the development of India, partnering industry, government, and civil society through advisory and consultative processes. The focus is on four key enablers: facilitating growth and competitiveness, promoting infrastructure investments, developing human capital and encouraging social development. The Climate Council of the CII was formed to strategize on implementation of the National Action Plan on Climate Change and to engage industry, policymakers and research and development (R&D) institutes to formulate strategies to commit to accelerate deployment of clean energy technologies, build capacity to access and internalize cutting edge technologies. The CII Green Services Division operates through the Green Business Centre (CII-GBC), offering niche green services to Indian industry. The objective of the CII-GBC is to promote green concepts leading to sustainable development, efficiency and equitable growth. Services offered include green process certification, green building certification (advisory services on construction of green buildings and award of a Green Building certificate), technology centres, training programmes on green-related topics, business incubation and facilitating entrepreneurs in developing and marketing new and innovative green products for commercialization. The CII Green Business Council has launched the following rating systems: IGBC Green New Buildings Rating System, IGBC Green Schools Rating System and the IGBC Green Mass Rapid Transit System.

None of the business and employers' organizations have any stated policy on climate change or any apparent interest in the impacts of climate change on the labour market. The strategies of the associations, on the other hand, have shifted away from government advocacy on behalf of business to cooperating with governments as an ally. Interestingly, the World Economic Forum in 2011 ceased its partnership of 28 years with CII to conduct its very influential annual event, the implication being that the WEF no longer saw the CII as bringing to the table sufficient influence with government. This is not necessarily a negative, it merely conveys that the priorities of the association are either shifting away from direct lobbying and advocacy as a core activity and indicator of effectiveness or they seek other forms and means of influencing government (there is probably a bit of both in effect).

The FICCI has a higher profile and stronger policy advocacy focus in its work on climate change, notable from the event arranged by its Climate Change Consultative Forum in June 2015 allowing members to engage with the Ministry working on the development of the Indian INDC to the government. The Forum was established for member exchange, policy development and advocacy, and hosted the Climate Change Enclave in March 2016, a repeat of the major event of 2013. The FICCI is the Indian arm of the International Chamber of Commerce and is leveraging the Chambers climate change policy work that is thoroughly canvassed across the many country members and is a consensus therefore representative view of the demands of industry. While the ICC has produced quality support material, publications and advocacy strategies, the FICCI has done little further domestic research or report production.

In sum, the FICCI is the stronger advocate of the two peak bodies. Although its effectiveness in influencing government is yet to be established, it is engaging the membership and therefore prima facie establishing an ecological consciousness. Combined, the FICCI and the CII are through their different approaches influencing behaviour and have considerable penetration, although they are neither addressing the labour market impacts of climate change nor are they either informing or pressing the government to consider the impacts.

The three Indian trade unions profiled in this study are Confederation of Free Trade Unions of India (CFTUI), Indian National Trade Union Congress (INTUC), and All India Trade Union Congress (AITUC) and are very similar in their objectives and structure. A feature of the Indian trade union movement is the volume of unions and, while membership numbers are large (in the millions), they are still a small percentage of the workforce. As well, they are generally known to be the trade union branch of a political party, with the AITUC affiliated with the Communist Party and the INTUC with the Indian National Congress. The Confederation of Free Trade Unions of India (CFTUI) boasts of its non-political affiliation and is emerging as a new force organizing those that are presently overlooked by trade unions and bringing together unions as affiliates.

CFTUI's membership is 321 unions and 1.1 million members. CFTUI is an umbrella organization of trade unions, trade federations, associations, societies and other organizations of India, dedicated towards welfare and development of the society in general and workers in particular. It is free from any control or attachment with the political parties or groups. The union does not engage with climate change issues.

INTUC has 33.3 million members, 29 branches and 26 industrial federations. Its main objectives are to establish an order of society that is free from hindrance in the way, of an all-round development of its individual members, which fosters the growth of human personality in all its aspects. It works to progressively eliminate social political or economic exploitation and inequality, the profit motive in the economic activity and organization of society and the anti-social concentration in any form. Further objectives are to place industry under national ownership and control in suitable form in order to realize the aforesaid objectives in the quickest time; and to organize society in such a manner as to ensure full employment and the best utilization of its manpower and other resources. The union does not engage in climate change issues.

AITUC's membership is 3.6 million. The unions affiliated to AITUC are from textile, engineering, coal, steel, road transport, electricity board and unorganized sectors such as beedi, construction and head-load workers, anganwadi, local bodies and handloom. The main objectives of AITUC are to establish a socialist state in India, to socialize and nationalize the means of production, distribution and exchange as far as possible and to ameliorate the economic and social conditions of the working class. The union does not engage with climate change issues.

The issues confronting many workers in India are basic rights and reasonable reward and these are therefore the focus of the trade union attention. The issues of this research project are therefore subjects they rely on their peak representative organizations (ITUC) to address and advocate. Each of the unions is active in the ITUC, which is a strong and informed advocate with policy developed and debated by its membership. It has not been possible to establish whether the Indian trade unions have a policy or opinion about climate change that they advocate at either the ITUC or in domestic political affiliations.

While the business associations are active in climate change in a limited way, influence in policy rests with the CEOs of the major members who are invited by government to consult and who rely on their international affiliations for policy direction and research. As employers' organizations, it is only the CII that has acted and then only in a limited way to address the labour market issues of planning or skill development. Trade unions have not engaged at all on climate change and the labour market impacts, their resources being committed to the first order issues of rights, wages and conditions. They are active in the international affiliates and engagement on climate change would be through the ITUC.

The government has made a solid commitment to emission reduction but it is constrained by economic, social and environmental factors. In an ecological modernization context, while it appears as a strong model of EM, it suffers from weak institutions of the state, external assistance is required to implement its plans to deliver electricity to all its population, the capacity of industry to reduce its fossil fuel dependence is limited and its engagement with civil society is also limited. The communities are aware of the need for adaptation and mitigation and support the change in patterns of behaviour but its capability is limited.

In the event that external funding to support the implementation of its INDC is not forthcoming, India will remain a major emitter of GHG's. This conditional commitment to the ecological outcome reflects a suboptimal model of EM and while it still informs the gaps in the policy process, the capacity to address those gaps is beyond the capacity of the state, industry and society to adequately respond.

KENYA

Kenya is a large country with the greater part of its land mass as arid and semi-arid land. These areas support almost 30 per cent of the population and 70 per cent of the livestock production (MENR, 2015). Eighty per cent of the Kenyan population works in the agricultural sector and over 75 per cent of the country's agricultural output is from small scale rain-fed farming or livestock production (CIA, 2016). Kenya is a low middle income country and faster growth and poverty reduction is hampered by corruption, reliance on primary goods and inadequate infrastructure. Chronic budget deficits have plagued the government's ability to implement proposed development projects (CIA, 2016).

Climate change is an important issue for Kenya. It is extremely susceptible to climate-related events that pose a serious threat to the socio-economic development of the country. Droughts and floods are having devastating consequences and, according to scientific evidence, are likely to continue to affect the country into the future (MENR, 2015). In many areas rainfall has become irregular and unpredictable and extreme and harsh weather is now the norm; for example, some regions are experiencing drought during the long rainy season while others are experiencing severe floods during the short rains. A reduction in cold extremes has been observed over the arid and semi-arid lands regions.

The adverse impacts are compounded by local environmental degradation, primarily caused by habitat loss and conversions, pollution, deforestation and overgrazing. Forest cover, for example, has reduced from 12 per cent in the 1960s to 6 per cent today (MENR, 2015).

Kenya's constitution requires the attainment of ecologically sustainable development. This requirement forms the basis for its climate change policy, efforts which began in the 1990s. In 2010 the Ministry of Environment and Natural Resources (MENR) launched the National Climate Change Response Strategy and later the 2013–2017 Climate Change Action Plan. The Strategy is a model of strong ecological modernization, identifying measures that include carbon markets, green energy development, research and development, an institutional framework for governance and community engagement.

Over the five years to 2015, the MENR has produced several drafts of the National Environment Policy that recognizes climate change as one of the direct causes of natural disasters and proposes measures to address climate change. This also formed the basis of the government's Climate Change Bill that, although approved by parliament in 2012, was vetoed by the President who cited lack of public involvement in the discussion of the new bill. The bill forms the National Climate Change Council and among other things also forms the Climate Change Fund and mandates public consultation for all climate change-related policy processes (LSE Grantham, 2015).

Interestingly, Kenya's energy requirements are met from biomass, which provides 69 per cent with petroleum (22 per cent) and electricity (9 per cent) supplying the remainder. Renewable energy sources contribute 74.5 per cent of electricity production, with hydro accounting for 50 per cent of the electricity production (LSE Grantham, 2015).

The main climate change legislative and policy instruments are the Energy Act 2006, parts of which are executed by the Energy Management Regulations 2012, National Environment Policy 2013, National Climate Change Response Strategy 2010 as implemented by 2013–2017 Climate Change Action Plan and the National Policy for Disaster Management. The main elements of the governments INDC submitted to the UNFCCC in 2015 are shown in Table 4.5.

Information about the employers' organization, business associations or trade unions is difficult to obtain, as their websites and other publicly

Table 4.5 Main elements of the INDC submitted to the UNFCCC in 2015

	National circumstance	Contribution	Contribution mitigation	Contribution adaptation	Fairness	Review	Means to implement
Kenya	80% arid. 75% GHG from land use	Reduce GHG emissions 30% by 2030 relative to BAU* subject to international assistance	Expand geothermal, solar and wind Tree cover of 10% Low carbon transport Climate smart agriculture	Mainstream adaptation into development plans subject to international finance and technology assistance	Poverty alleviation and economic development must be considered	Every 5 years	US$40 billion by 2030 from domestic and international sources

Note: *BAU scenario of $143MtCO_2$.

available sources such as government registers (for statutory documents), media reports and publications do not provide information from which an organizational profile can be reasonably developed.

There is little to report from either the employers' organization Federation of Kenya Employers (FKE) or the business association Kenya National Chamber of Commerce and Industry (KNCCI) in regard to climate change. The FKE is a large organization but its sphere of interest is limited to direct industrial relations and does not extend to broader labour market planning or other policy issues such as climate change. The KNCCI similarly has a narrow focus of trade and business activity.

FKE membership is 4,000 employers and includes employers in the private and public sectors including state corporations, the local authorities and employers' associations. The objectives of FKE are to act as a forum for employers, to promote and defend the interests of employers, to promote good management practices, to collaborate with employers, intergovernmental and other business organizations and to develop a sustainable institutional capacity and competence. It does not appear that the organization is directly engaged on climate change issues.

The trade union, the Central Organization of Trade Unions (COTU) does not have an active climate change advocacy or member information programme. Its focus is on organizing the workforce and advocating the rights of its members on workplace issues. COTU membership is 41 trade unions. Its principle objectives are to improve the economic and social conditions of all workers in all parts of Kenya, to assist in the complete organization of all workers in the trade union movement and to organize the structure and spheres of influence and amalgamation of trade unions affiliated to COTU.

In Kenya, employers' organizations and trade unions are not engaged in the climate change policy development process. The broader business community also does not engage in this process. This is a reflection of the organizations' focus rather than any apparent barriers to engagement created by the government or departments that are mandated to engage with the community and stakeholders.

The constitutional requirement for ecologically sustainable development forms the basis for its climate change policy, efforts which began in the 1990s. In 2010 the Ministry of Environment and Natural Resources (MENR) launched the National Climate Change Response Strategy and later the 2013–2017 Climate Change Action Plan. The Climate Change Bill, which forms the National Climate Change Council and among other things also forms the Climate Change Fund, mandates public consultation for all climate change related policy processes. The Strategy contributes to an optimum ecological modernization model for the country, identifying measures that include carbon markets, green energy development, research and development, an institutional framework for governance, and community engagement. The strength of the model is however mitigated by the stated dependence on external finance to implement the strategies, and the weakness of the institutions of the state, and of industry to innovate.

NOTES

1. There is no effective peak national employers' organization in Canada. Representation of Canada at the International Organisation of Employers is by a consultant engaged to monitor activities who, while a respected participant, has no mandate to advocate on behalf of Canadian employers.
2. The UN Global Compact is a voluntary commitment by signatory companies to align strategies and operations with universal principles on human rights, labour, environment and anti-corruption, and take actions that advance societal goals (UN Global Compact, 2016).

5. Case study: the European Union

INTRODUCTION

The EU has long been at the forefront of international efforts to combat climate change and was influential in the development of the United Nations' climate treaties: the 1992 UN Framework Convention on Climate Change (UNFCCC) and the 1997 Kyoto Protocol (EC, 2011a). In March 2007, EU heads of government agreed to reduce emissions by 20 per cent by 2020. In 2014, this figure was revised to 40 per cent by 2030. In the months following the 2007 decision, the European Commission released a detailed plan in line with the longer term perspective set out in the *Roadmap for moving to a competitive low carbon economy in 2050*, the *Energy roadmap 2050* and the *Transport white paper*. The EU had a single-minded focus on climate change and was prepared to embark on this costly programme without a similar commitment from other major contributors to global emissions such as the United States (US), China and India. Schmidt (2008) contends that European policy on climate change is a reflection of the European political and social culture, noting that most western Europeans have long been concerned about global warming and other environmental problems. This was demonstrated in the 1997 decision of the heads of member states to amend the Treaty establishing the European Community to embed protection of the environment as the Treaty's third pillar alongside economic growth and social protection (Europa, 2012b).

GOVERNING INSTRUMENTS

The constitution of the EU merits comment because of its unique statutory relationship with member states. The treaty that forms the EU, the Treaty of Maastricht, includes provisions for the protection of the environment and contains the requirement that the EU must consult on matters of social policy with employers, workers and civil society. Both are matters not usually addressed in the rules governing the state. Another feature of the EU model is the statutory framework that sees the coexistence of regional (European) law with the domestic law of the member states, a model replicated only in countries that are federations with the right to frame laws at federal and state levels such as the US, Australia and India.

The European Union is an economic and political union of 28 member states and was formed by a treaty process. The first treaty was written in 1951 as the European Coal and Steel Community and was followed by the 1957 Treaties of Rome, which established the European Economic Community and the European Atomic Energy Community. The 1993 Treaty of Maastricht formally established the European Union and the concept of European citizenship (Eupolitix, 2013). Since its initiation in 1951, membership has grown from six to 28 states at 2007 and the scope of the treaties has expanded. The 1997 amendment to the Maastricht Treaty added protection of the environment to the objectives of economic growth and full employment. The amendment provided that the European Community:

> shall work for the sustainable development of Europe based on balanced economic growth and price stability, a highly competitive social market economy, aiming at full employment and social progress, and a high level of protection and improvement of the quality of the environment. (European Community, 1999, Art. 3)

The belief that a post-war Europe should provide its citizens economic and social equity is reflected in the provision in the 1957 Treaties of Rome, which required that the representative organizations of business, workers and civil society be consulted in the process of developing European law and practice. The Treaty at Article 300.1 provides that 'the European Parliament, the Council and the Commission shall be assisted by an economic and social committee and a committee of the regions, exercising advisory function and . . . [the committee] shall consist of representatives of organisations of employers, of the employed and of other parties' representative of civil society' (European Union, 2010, Art. 300.1). To ensure this happens, the European Economic and Social Committee (EESC) was established, through which the European Parliament, Council and the Commission is required to consult on matters such as social policy, social and economic cohesion, environment, education, health and consumer protection.

The Treaties are binding agreements with member states that set out the objectives and rules for the EU institutions and how decisions are made. Under the Treaties, the EU institutions can adopt legislation on which member states are required to act. The EU cannot propose law that is outside the scope of those Treaties. The laws of the EU function alongside the individual laws of each member nation. When there is conflict between the EU member nations' law and EU law, EU law takes precedence (Eupolitix, 2013). The pervasiveness of EU law in the policy and practice of the state is further observed in the UK study in Chapter 6.

The EU and member states can also establish charters. Their legal status may be one of political declaration if the agreement that forms the Charter is not unanimous and where it is unanimous it forms part of the treaty and has legal status (European Parliament, 2001). Charters adopted by the member states include the Community Charter of the Fundamental Social Rights of Workers (Eurofound, 2011a), European Social Charter (Council of Europe, 1996) and the Charter of Fundamental Rights of the European Union (European Parliament, 2007).

The three pillars of economic, social and environmental wellbeing in the 1997 Treaty on the European Union and the suite of social charters are a reflection of the political and social expectations of European people. It is necessary to recognize these expectations in order to understand why employers' organizations and trade unions are afforded special and protected status through the EC Treaty and the Charters. The Treaty and the Charters integrate the social partners into the fabric of the institution and the social policy regime envisaged in the Maastricht Treaty reforms (Barnard, 2002; Falkner, 2011).

Eurofound (2011c) research concludes that social dialogue helps to enrich the process of governance in the EU by including the social partners as informed actors to participate in the deliberative process and that the social partners are core actors in that process and must have the right to participate. The qualification to this support for the social partners and the maintenance of the democratic element in the deliberation is that the social partners still need to be representative and to be internally democratic, transparent and accountable. Accordingly, more attention needs to be paid to ensure the quality of the deliberative process (Barnard, 2002; Ebbinghaus, 2002; Falkner, 2011). This does not fully answer the question of why the EU continues to place faith in organizations that underperform in some areas. However, it does lead to the conclusion that the EU's faith is a reaffirmation of the community's acceptance that employers' organizations and trade unions are rightful participants in the policy process. At the European level, the peak bodies that are the social partners have almost unanimous affiliation with the organizations in the member states. It is the state organizations where the representativity is challengeable.

The social partners in the EU member states are widely reported as being supportive and active participants in the process of policy development and delivery on sustainability and climate change (EC, 2011a). The European Commission views social partners as playing an important role in the economy as a whole and the labour market in particular. With respect to climate change, the EU looks to their social partners to create consensus on the implementation of policies across industry and society. The EU has confidence in the ability and leadership of the social partners, expressing the view that 'a shared analysis of employment opportunities and challenges by social partners can contribute greatly to a well-managed and socially just transition' to a low carbon economy (EC, 2011a, p. 153).

EU CLIMATE AND ENERGY COMMITMENT

In March 2007, the EU leaders adopted an integrated approach to climate and energy policy that aims to combat climate change and increase EU energy security while strengthening competitiveness. They committed to the 20:20:20 targets of a reduction in EU greenhouse gas emissions of at least 20 per cent below 1990 levels; 20 per cent of EU energy consumption to come from renewable resources; and a 20 per cent reduction in primary energy use compared with projected levels by improving energy efficiency, all to be achieved by 2020 (Europa, 2010b). The core package consists of four complementary programmes: revision and strengthening of the emission trading system (ETS); an effort-sharing rule governing emission from sectors not covered by the ETS; binding national targets for renewable energy; and a legal framework to promote the development and safe use of carbon capture and storage. The package creates pressure to improve energy efficiency but does not address it directly, which is done by the EU energy efficiency action plan (EC, 2006). Further sectoral initiatives address sustainable mobility (EC, 2011c) and the construction sector, and are listed in Table 5.1.

Table 5.1 Construction sector initiatives

Flagship initiatives	Legislation
• Resource efficient Europe • Low carbon economy 2050 roadmap • Roadmap for a resource efficient Europe • Energy 2050 roadmap • Energy efficiency plan 2020 • Strategy for the sustainable competitiveness of the EU construction sector • Innovation Union	• Energy Services Directive2006/32/EC • Eco Design Directive 2009/30/EC • Labelling Directive • EPBD 2010/31/EU • Construction Products Directive

Source: BPIE (2011).

In 2014, the EU adopted its 2030 Climate and Energy Framework, which succeeds the EU 2020 climate and energy package. It sets three key targets for the year 2030: at least 40 per cent cuts in GHG emissions from 1990 levels; at least 27 per cent share for renewable energy; and at least 27 per cent improvement in energy efficiency.

The framework is also in line with the longer-term perspective set out in the *Roadmap for moving to a competitive low carbon economy in 2050*, the *Energy roadmap 2050*, and the *Transport white paper*. In addressing GHG emissions, the framework provides for making fair and ambitious contribution to the new international climate agreement, the ETS sectors would have to cut emissions by 43 per cent and non-ETS sectors would need to cut emissions by 30 per cent. It is estimated an annual investment in emission reduction programmes of US$38 million is expected to be covered to a large extent by fuel savings (Europa, 2016).

The development of strategies to achieve the targets is the responsibility of the member states and the strategies adopted differ depending on the particular features of the domestic economy, society and stages of development. The Climate Policy Tracker for the EU (WWF, 2011) was a 2010 report of the findings from surveys of all EU member states using 83 indicators to measure policy effectiveness. It found that based on present policies and member state commitment, the EU targets will not be met and most states will only achieve one-third of their targets, with the exceptions of the major economies of France, Germany and the UK that have already met their Kyoto targets and were well on track to meet their aims under the EU Climate and Energy Package. In the end the EU did meet its Kyoto targets but the prediction of underperformance of some states held true. The Climate Policy Tracker also reported that the performance for many states is defined by the ambition in EU legislation and that EU policies assist many states in formulating policy. This is not surprising, but it does provide a further insight into the pervasiveness of EU legislation and policy in the behaviour of member states.

SOCIAL PARTNERS, EMPLOYER ORGANIZATIONS AND TRADE UNIONS

'Social partners' is a term generally used to refer to representatives of management and labour. 'European social partners' specifically refers to those organizations at the EU level that are engaged in European social dialogue as provided under Articles 154 and 155 of the Treaty on the Functioning of the European Union (Eurofound, 2011c).

European law recognizes as social partners the peak organizations of organizations which themselves are an integral and recognized part of member state's social partner structures and have the capacity to negotiate agreements (Eurofound, 2011c). The major and representative European social partner organizations are BusinessEurope and the European Trade Union Confederation (ETUC).

The advocacy of BusinessEurope and business and employers' organizations in member states concerning climate change is aimed at ensuring companies remain competitive during the greening of the economy process (BusinessEurope, 2013b). Their perspective of a low carbon economy is one that is framed around cost-effective policy options, investment in infrastructure, promoting green technology, ensuring a level playing field internationally, minimalist regulation and corporate social responsibility options (BusinessEurope, 2013c). Activities by employers' organizations on greening the economy are generally directed at engaging in political debates, publishing position papers and promoting green skills (Eurofound, 2015).

ETUC and trade unions in member states advocate for a just transition that includes dialogue, skill adaptation and investment in green jobs (ETUC, 2010). Activities by trade unions on greening the economy are generally directed towards demanding a voice as a political actor. Their instruments are involvement in political debates, publishing position papers, conducting training for union representatives and initiating actions in individual workplaces (Eurofound, 2015).

EU research reports that social partners in the member states are involved in policymaking on low carbon economy issues from policy formulation onwards through to implementation, where they express their positions on policy proposals and programmes (Eurofound, 2011b; EC, 2011a). At the EU level, the ETUC is active at the stage of policy formulation but this engagement is not shared by BusinessEurope, whose regular contribution is as a participant in peer-review panels for proposals generated by the European Commission or Parliament.

BusinessEurope is a federation of national business and industry associations in Europe. The organization's membership is made up of 41 national federations from 35 countries (BusinessEurope, 2013d). It does not have sectoral associations or corporate or individual members, although it does have 55 partner companies that may access its networks within the EU and participate in its working groups (BusinessEurope, 2013d). The only formal alliances maintained by BusinessEurope are with the European Commission through the EESC, and the NGOs Alliance for a Competitive Europe and Alliance for CSR. It has informal affiliations with groups formed for specific purposes such as the Climate Change Roundtable. BusinessEurope members are individually affiliated with international agencies such as the International Labour Organization (ILO) or the

International Chamber of Commerce (ICC) and BusinessEurope only participates in the international climate and sustainability events as part of the Euro delegation.

The governing body of BusinessEurope is the Council of the Presidents, which determines its general strategy. The Executive Bureau comprises representatives of 'the five largest countries, the country holding the EU presidency and five smaller countries on a rotation' (BusinessEurope, 2013d, p. 1). The Bureau monitors the implementation of the annual programme, keeps other member organizations informed of progress and responds to issues as they arise. The Executive Committee is a committee of the Directors-General of all members who are responsible for translating strategy into activities and tasks. There are seven policy committees and approximately 50 working groups supported by a secretariat of 45 staff based in Brussels (BusinessEurope, 2013c).

BusinessEurope describes its main task as ensuring that company interests are represented and defended vis-à-vis the European institutions with the principal aim of preserving and strengthening corporate competitiveness (BusinessEurope, 2013c). BusinessEurope explains that it is actively engaged in European social dialogue in order to find solutions reconciling the economic and social needs of labour market players and to devise concrete arrangements that benefit both companies and employees. The BusinessEurope brief for employment and social affairs responds to the concerns about rising unemployment, an ageing workforce and increased international competition. It contends that structural reforms are needed to improve labour market flexibility, secure the availability of a skilled workforce – including through economic migration – and put in place modern social policies. Their aim is to have more people in work, working more productively (BusinessEurope, 2013e).

BusinessEurope is credited as always giving detailed input in all aspects of the EU climate and energy policy at the political and technical levels. It broadly supports the EU climate change objective, but insists that industry's international competitiveness and energy security should not be harmed by unilateral EU action (EC, 2011a). Until recently, BusinessEurope had not established a profile as an opinionated advocate on environmental and related regulatory issues. Its advocacy on climate change policy was concerned with energy security and ensuring companies remain competitive in the process of greening the economy (Eurofound, 2011a; BusinessEurope, 2013b). However, during 2013 BusinessEurope campaigned vigorously against the 2012 European Commission proposal for a twin tranche approach to fixing the depressed carbon market by short-term back-loading (withholding the release of the next tranche) of carbon allowances, to be followed by long-term structural change before the end of 2013 (Euractiv, 2012). BusinessEurope's stance divided industry, with BusinessEurope lobbying hard against any interference in the market and, in opposition, a coalition of energy companies and others welcoming the initiative and calling for it to be extended further (Euractiv, 2012).

The EU's intervention proposal was ultimately defeated in the parliament. It was reported in the EU monitor Euractiv (2013) that BusinessEurope's lobbying had been strongly influential in the vote and that a letter sent by BusinessEurope's Director to the EU President Herman Van Rompuy 'showed that the industry group priorities had influenced the [Parliament's] agenda' (p. 1). The subsequent release of the European Commission Green Paper *A 2030 framework for climate and energy policies* has seen BusinessEurope maintain its vocal advocacy, arguing the review should address the flaws in the package rather than proposing further intervention (BusinessEurope, 2013f).

In the lead up to the 2015 Paris COP, BusinessEurope advocated for an ambitious legally binding global agreement, which reflects the long-term objective of limiting global warming below 2 degrees Celsius. Its platform was that the development of a global carbon market should play a stronger role in the future and economic instruments can best help to stimulate investment in innovative low-carbon technologies and products in locations where they deliver the greatest. It fully endorsed the EU Emissions Trading Scheme (ETS) as the cornerstone of EU climate policy, while keeping strong protection measures for sectors on the global industrial market until main competitors have comparable carbon costs.

The ETUC (2013b) describes its purpose as speaking with a single voice on behalf of the common interests of workers at European level. Its prime objective is to 'promote the European Social Model . . . where working people and their families can enjoy full human and civil rights and high living standards' (ETUC, 2013b, p. 1). Membership of the ETUC consists of 85 national trade union confederations from EU member states and 10 European trade union federations (ETUC, 2013c). The ETUC Congress meets every four years and is responsible for the overall policy and decision-making of the confederation. The Executive Committee and smaller steering committees are responsible for implementing policy between congresses, while the Brussels-based Secretariat runs the day-to-day activities (ETUC, 2013b).

The ETUC made the issue of climate change a priority of its sustainable development strategy in 2002 (Eurofound, 2011a). It drew up the first 'Union proposal for a European policy on climate change' (ETUC, 2004) and actively debated proposed EU climate change legislation including the green paper on energy efficiency (ETUC, 2011a), the revision of the EU's emissions trading directive and the Climate and Energy Package (ETUC, 2011a). It established a civil society coalition in 2001 with the NGOs

European Environmental Platform and Platform of European Social NGOs, which annually submit recommendations for a social and sustainable Europe to the EU Councils Spring Summit (EC, 2011a).

The ETUC also commissions research both through the union-sponsored research agency and think tank European Trade Union Institute and independently. Their 2007 research *Climate change and employment: Impact on employment in the EU-25 of climate change and CO_2 emission reduction measures by 2030* (ETUC, 2007) was a major contributor to the debate concerning the impact of climate change policy on the labour market. It evaluates the economic shifts and employment consequences of climate change and serves as a useful companion to the ILO's report on the impact of jobs in the transition to a low carbon economy (Worldwatch Institute, 2008). The major recommendations from that research are that planning for policies that have a social impact should be through tripartite forums and involve social dialogue and collective bargaining; policies should be easier to predict to allow the anticipation of the social consequences and ruptures; and measures should be taken to maximize positive spin-offs for employment and setting socially sustainable criteria for publicly funded projects (ETUC, 2007).

The ETUC approach to policy development and member engagement to achieve leverage in the EU is demonstrated by its 2013 sustainable mobility research project (ETUC, 2013d). The research process brought together industry stakeholders in a series of workshops and commissioned research externally to inform the policy development on sustainability in the transport sector. The ETUC's advocacy is based on what it terms the five pillars of a just transition: stakeholder consultation, green and decent jobs, the responsibility of the state, labour rights and social protection. This is a more expansive definition than the ILO definition of a just transition which is based on the ILO Labour Conventions and provides for the recognition of workers' rights, decent work, social protection and social dialogue (Worldwatch Institute, 2008).

The ETUC's current strategy is reflected in its ETUC action programme 2015–2019 Re: Sustainable Development, in which it declares it will pursue the following objectives:

- A change to the European and global economic model based on long-term investment, a stable but ambitious regulatory framework and a strong social dimension so as to bring about a 'just transition' to a green economy for all Europeans.
- A sustainable investment strategy for Europe.
- No funding for projects at odds with the environmental commitments of the EU.

- Development of a low-carbon and sustainable strategy for European industrial policies.
- A just transition policy framework with strong EU financial support based on the five pillars of social dialogue, investment in quality jobs, greening of education, training and skills, trade union rights and social protection to tackle climate change (both mitigation and adaptation) at both the European and international levels.
- An effective European energy community.
- A resource-efficient Europe.
- The greening of the labour market.

CONCLUSION

The EU and its member states recognize the necessity of the transition to a low carbon economy and that this transition involves social and economic opportunities and costs. The state is the main actor in the transition, although the social partners in the EU are regarded as important stakeholders and social dialogue has an important role to play, helping to create consensus among membership for climate-related policies.

The essential difference between the ETUC and the BusinessEurope approaches is that the ETUC is actively advocating its position and flooding the discussions about social policy and climate change with research and policy papers that support its position. BusinessEurope, in contrast, is more issues focused, responding to current issues in order to draw attention to its policy position. The European Commission (EC, 2011a) rationalizes these different approaches, noting that social partners act first and foremost where they have direct competencies. For the ETUC, this is the distribution of benefits, rights and obligations of workers, while for BusinessEurope it is how the transition to a low carbon environment will impact on the economy.

This chapter provides a number of significant contributions to the book. It explores why and how employers' organizations and trade unions are afforded special and protected status and integrated into the fabric of the European institution. It poses the question of why the EU continues to place faith in organizations that underperform in some areas and where their legitimacy and representativity is challenged. It concludes that the EU and the member states as the statutory authorities are the main actors in the transition to a low carbon economy, and the social partners are rightful participants in the policy process and provide a supporting role.

6. Case study: United Kingdom

INTRODUCTION

Overview

Two detailed studies were undertaken from among the countries and the region that were profiled for this book, the European Union and United Kingdom. The European context was chosen because Europe and its member states present a rich source of policy and experience on climate change. Europe has been at the leading edge of developments in climate change and the EU and member states have been taking action since the early 1990s to achieve reductions in GHG emissions. They have also worked actively for global agreement on climate change. The UK was chosen because it is a mature industrialized economy with a tradition of intervention by the state in environmental management and, through the 2050 Pathway Package (DECC, 2010), has set in place a programme supported by legislation and finance to ensure progress towards this future commitment.

The UK's Ecological Objectives

The 1997 Marshall Report (Dresner et al., 2006) discussed the social responses to environmental tax reform in the UK. Evidence of the UK's ecological objectives associated with climate change and energy security can be seen in policy documents released since then, which have variously engaged with and addressed these issues. The UK government's commitment to GHG emission reduction and its approach to the issue of climate change are reflected in a number of formal instruments across international, regional and domestic statutory jurisdictions (Table 6.1). The UK has made significant progress towards these commitments and its emissions reduction required under the Kyoto Protocol have been achieved (Syndex, 2011). However, the Committee on Climate Change (CCC), established under the UK Climate Change Act 2008, has continued to warn that the longer term commitments to reduced emissions by 80 per cent below 1990 levels by 2050 may not be achieved without further intervention by the government because the current programmes are failing to achieve the necessary step

Table 6.1 UK's formal commitments to greenhouse gas emission reduction

Commitment type	Commitment
Domestic commitments	• 2050 Pathway Analysis (DECC, 2010) • Climate Change Act 2008: Emissions reduced by 80% below 1990 levels by 2050 (Carbon) budgetary period including the year 2020 • Carbon Plan (DECC, 2011a) • Climate Change Act 2008 • Energy Act 2008
International commitments	• UNFCCC Agreements from the COPs • KP: Emissions to be reduced by 5% below 1990 levels to be achieved during 2008–2012 (UNFCCC, 1997)
Commitments to the European Parliament	• European Union (EU) Climate and Energy Framework: 40% reduction in greenhouse gas emissions, 27% of energy from renewable sources and 27% reduction in primary energy usage, by 2030 (Europa, 2016) • Greenhouse Gas Effort Sharing Decision: Emissions reduced in the sectors of the domestic economy not covered by the EU Emissions Trading System (ETS) which in the UK is a reduction in non-ETS sectors equivalent to 16% below 2005 levels (DECC, 2011c)

changes (DECC, 2011e; 2014; 2015). The Committee on Climate Change was formed to advise the UK government and devolved administrations on emissions targets and report to parliament on progress made in reducing GHG emissions and preparing for climate change. The CCC provides independent advice to the government on setting and meeting carbon budgets and preparing for climate change, monitors progress in reducing emissions and achieving carbon budgets, conducts independent analysis into climate change science, economics and policy and engages with a wide range of organizations and individuals to share evidence and analysis.

PUBLIC POLICY

Policy Framework

In the long history of the UK, there have been a number of significant environmental disasters that occurred as the result of human activity and required intervention by the state to be resolved. Examples include the 1862

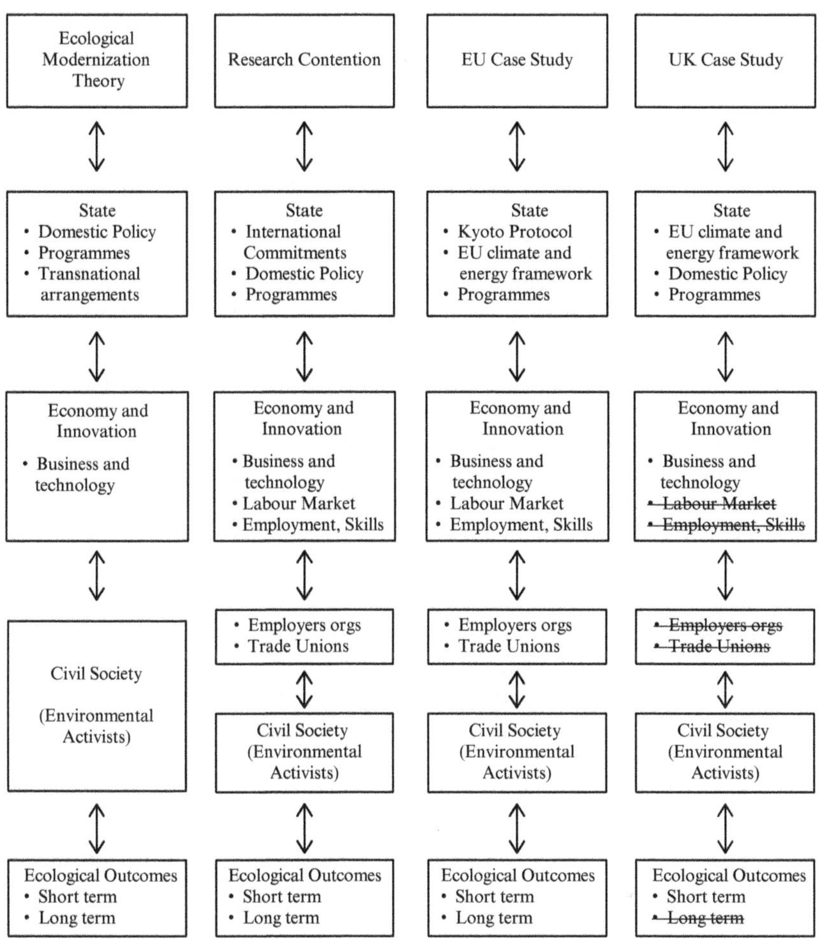

Figure 6.1 Ecological modernization, the research contention, and the EU and UK units of the study

Royal Commission on River Pollution, which was initiated in response to the typhoid and cholera epidemics of the 1840s (Radford University, 2012); the 1956 Clean Air Act, which occurred in response to the Great Smog of 1952 where weather conditions and air pollution combined to result in a layer of thick fog forming over London, causing ongoing damage to the health of the city's residents (University of Edinburgh, 2012); and the 1972 Royal Commission on Environmental Pollution, which was established to address the environmental problems and pollution caused by road traffic (Dresner et al., 2006).

A more structured approach to management of the environment and climate change by the state in the UK began when climate change emerged as a dedicated priority issue at the 1992 Rio Earth Summit and the 1997 UNFCCC COP adopted the Kyoto Protocol which saw 37 industrialized countries and the European Community, including the UK and known as the Annex 1 Parties, formally commit to reduce GHG emissions to an average of 5 per cent against 1990. The UK parliament ratified its commitment in 2002 (UNFCCC, 1997; UK POST, 2002).

In 2001, the UK government legislated to impose a levy on industrial and commercial activity as a means of reducing GHG emissions. This was followed by the introduction of an emission trading system (ETS) in 2002 (Dresner et al., 2006). In 2008, the Climate Change Act that set a target for the reduction of GHG emissions of 80 per cent below the 1990 base year by 2050 was passed into law (DECC, 2011b).

A chronology of the major developments concerning the UK's climate change policy until 2015 is outlined in Table 6.2.

Much of the domestic climate policy of EU member states was initially framed by requirements of the EU's 2020 Climate and Energy Package and related regulations intended to combat climate change and increase the EU's energy security (Europa, 2010b). At the 2005 Hampton Court Summit, an informal meeting of the European Council convened by the Council President to discuss Social Europe, innovation and globalization issues, EU member states gave the European Commission a mandate to develop a common energy policy (Europa, 2010b). In March 2007, the European Council 'approved the ambitious climate change and energy package to build a low carbon economy in Europe' (DECC, 2007, p. 34). The targets, to be met by 2020, were a reduction of EU GHG emissions of at least 20 per cent below 1990 levels, 20 per cent of EU energy consumption to come from renewable resources and a 20 per cent reduction in primary energy use compared with projected levels, to be achieved by improving energy efficiency (Europa, 2010b). The package became European law in 2009 and the targets became legally binding for member states. The EU's Lisbon Strategy for Jobs and Growth, which sought to have the EU 'become the most dynamic and competitive knowledge-based economy in the world, capable of sustainable economic growth with more and better jobs and greater social cohesion and respect for the environment' (BIS, 2008, p. 11), provided the bridge between the EU 2020 Climate and Energy Package and the subsequent 2030 Climate and Energy Framework, and its policy for labour market and skills. The strategy set the aim of achieving 'sustainable economic growth with more and better jobs and greater social cohesion' (p. 31), objectives the UK government endorsed.

In its *Green growth strategy synthesis report*, the OECD (2011) provides an

Table 6.2 Chronology of the major developments regarding climate change in the UK until 2011

Date	Development
1988	Contribution by UK scientists to Intergovernmental Panel on Climate Change (IPCC) first report
1992	Rio Earth Summit
1997	KP adopted
1998	Marshall Report proposed a climate change levy (CCL) as an economic instrument to reduce the industrial and commercial sectors greenhouse gas (GHG) emissions
2000	UK Pension Act amended to require trustees to declare social, environment and ethical considerations in investment evaluations
	Carbon Disclosure Project: fund managers sought information from the 500 largest companies in the world on their GHG emissions
2001	Climate Change Levy (CCL) introduced
2002	UK Emissions Trading System (voluntary) commences[1]
	Renewables Obligation: UK electricity suppliers were required to source an increasing proportion of electricity they supply to customers from renewable sources
2003	UK Energy White Paper: prescribed a long term strategic vision for energy policy combining environment, security of supply, competitiveness and social goals
	European Parliament adopts the EU Emission Allowance Trading Directive
	World Wide Fund for Nature releases its Powerswitch Financial Report
2005	EU ETS commences
	Friends of the Earth launch the 'Big Ask' campaign
2007	European Council approves the EU 20:20:20 by 2020 Climate and Energy Package
	Climate Change Bill introduced to parliament
2008	Climate Change Act passed into law
	UK Nuclear White Paper: new nuclear power stations should have a role to play in the UK future energy mix, alongside other low carbon sources
	UK Energy Act 2008 passed into law
	EU Energy Performance in Buildings Directive prescribed the methodology and standards for calculating and integrating energy performance in buildings
2010	UK Energy Act amended to cover Carbon Capture and Storage (CCS) demonstration projects, incentives and fairness of energy markets
2011	UK Energy Act amended to introduce the Green Deal
2015	Department announced it would not provide any further funding for Green Deal loans and the energy company obligation will end on March 2017, effectively bringing the schemes to a halt

explanation of why labour market planning is an important component of the strategy to transition to a low carbon economy and workplace. The report discusses the employment and distributional effect on the labour market of the impacts of climate change policy, noting that some jobs will be at risk and others will have to be reallocated, a churn that will create challenges for some and opportunities for others. It contends that labour and skills policies are important and that labour policies should focus on preserving employment rather than jobs. The policies need to ensure that workers and companies are able to adapt quickly to changes brought about by the greening of the economy, including by seizing new opportunities. It speaks of the need to reinforce social protection systems for those who are negatively affected.

The UK 2050 Pathway Analysis (DECC, 2010), the Carbon Plan (DECC, 2011a), the Climate Change Act 2008 and the Energy Act 2008 represent the UK's vision and plans for reducing emissions, climate management and energy efficiency. The long-term perspective offers some certainty for decision-makers against the potential short-termism in policy and facilitates regulatory coherence. The OECD's (2010) report *United Kingdom policies for a sustainable recovery* and the WWF (2011) *EU climate policy tracker* generally agree that UK policies and programmes are appropriate to ensure its commitments under the EU Climate and Energy Package can be met. These initiatives are considerate of the need for industry to remain competitive and to innovate. However, they do not explicitly integrate industry policy, the social dimension of climate change and the impact on the labour market, employment and skills requirements. Nor do they reference the principles of engagement with civil society and social dialogue that are so much a part of EU practice.

The 2050 Pathway Analysis

The UK 2050 Pathway Analysis (DECC, 2010) is rooted in scientific and engineering realities, looking at what is thought to be physically and technically possible (DECC, 2012a). It is system wide, covering all parts of the economy and all GHG emissions released in the UK. The framework for action that responds to the Analysis seeks to balance the UK's ongoing dependence on fossil fuels for energy and the targeted reductions in GHG emissions. It addresses the requirements for ambitious per capita reductions in energy demand; substantial increase in the electrification of heating, transport and industry; increased supplies of electricity; sustainable bioenergy; development in carbon capture and storage (CCS) technology; and managing emissions from agriculture, waste and transport (DECC, 2012a). The Analysis is recognition that the successful shift to a low carbon economy requires a clear direction and early action by the UK government to provide the confidence consumers and investors require for investment in infrastructure and plans that often have long payback periods.

The Carbon Plan 2011

The Carbon Plan (DECC, 2011a) is the second pillar of the UK government's climate change commitments. It is effectively a work plan that sets out how the UK will make the shift to a low carbon economy 'while maintaining energy security, and minimizing costs to consumers' (DECC, 2011a, p. 1). It was presented to parliament on 1 December 2011 and lays out a detailed programme of action with the timelines necessary to meet its objectives within the four budget cycles of the Plan. The sectoral plans cover buildings, transport, industry, electricity, agriculture, land use, forestry and waste and are required to address the related sustainability issues including sustainable low carbon energy, saving energy in homes and communities, reducing emissions from business and industry, moving towards low carbon transport, cutting emissions from waste, managing land sustainability, reducing emissions in the public sector and developing leadership within the EU (DECC, 2011a, Part 2). The Plan acknowledges the uncertainties of planning for emissions reduction such as how far demand can be reduced, the availability of biomass, how far electrification can be taken and which of CCS or nuclear will be the most cost-effective option (DECC, 2011a, Part 3).

The UK government's modelling finds that the country is in a strong position to deliver on its carbon budgets out to 2020 (DECC, 2011a), noting that achievement of the fourth budget to 2027 will depend on a range of variables such as economic growth and the related demand for energy, the EU ETS cap,[2] the ability to scale up the current strategies and the development of technology options such as carbon capture and storage. Despite the optimism the Plan's scenario modelling has generated, assessments by others including the government's Committee on Climate Change are less positive, finding that these measures are 'not likely to be sufficient to deliver on the fourth carbon budget' (DECC, 2015).

The Climate Change Act 2008

The Climate Change Act 2008 converts into law the ambition of the 2050 Pathway Analysis and the programmes of the Carbon Plan. The Act provides for the system of carbon budgeting, establishes a Committee on Climate Change and confers powers to establish trading schemes for the purpose of managing emissions. The Act ensures data collection is timely and reported regularly to provide information from which both the public and private sectors can make informed decisions.

The Act requires the Committee on Climate Change to report annually to the government with regard to adaptation as well as mitigation, and for this purpose the Committee 'established the Adaptation Sub Committee . . . to provide independent and expert advice on how to assess climate risks and to report regularly on the UK's progress in preparing for the future climate' (DECC, 2011b, p. 8). Its annual reports have consistently found that while the UK is coping with climate change, generally climate risks appear not to be fully incorporated into some major strategic decisions such as land use planning and investment in water infrastructure. They have also reported limited evidence of action by householders to reduce their vulnerability to unaffordable energy costs, which they see as indicating the need for new policy measures (DECC, 2015).

The Energy Act 2008

The Energy Act 2008 implements the legislative aspects of the Energy White Paper 2007 and the EU Climate and Energy Package (Europa, 2010a). The 2010 and 2011 amendments provide for the Green Deal (DECC, 2012b) initiative and the step changes required for the energy-efficiency measures to homes and businesses. They also make improvements to the regulatory framework for low carbon energy initiatives and fair competition in energy markets (DECC, 2011d).

The UK government's scenario analyses (DECC, 2007) predict that the UK will have an ongoing dependence on fossil fuels for energy. Accordingly, nuclear power and CCS technology are seen to be necessary elements of an energy policy as they assist with emissions management (DECC, 2010). They also indicate the UK government and society are less philosophically opposed than other nations to nuclear power. The government proposed the construction of four CCS demonstration plants in the period 2010 to 2020 that it estimated had the potential to create 8,000 jobs immediately and with an estimated employment potential of 30,000 jobs. However, 2015 saw the cancellation of the Commercialisation Programme for carbon capture and storage (CCS), accompanied by a reduction in funding for energy efficiency and cancellation of the zero carbon homes standard (UK CCC, 2016).

ENERGY EFFICIENCY AND THE GREEN DEAL

The EU 2020 Climate and Energy Package (Europa, 2010a) and the UK Energy Act (2008) mandated energy efficiency as a core element of the initiatives to build a low carbon economy. The UK planned initiatives to address this mandate were the introduction of smart meters and the Green

Deal, programmes that were to come into effect in 2014. As domestic buildings are responsible for 25 per cent of the emissions in the UK and just over 40 per cent of its final energy use (WEF, 2011), buildings are a logical initial focus. The Green Deal and smart meters were to be the flagship initiatives designed to improve the energy efficiency of UK building stock. The government intended that all new homes would be zero carbon from 2016 and non-domestic buildings from 2019 (DECC, 2011a).

Smart meters were to be mandatory in the 30 million homes and small businesses in the UK from 2019 and the information they provide should enable consumers to better understand their energy use and maximize the opportunities for saving energy (DECC, 2011a). The government would include energy-efficiency advice as part of the installation visit and provide support through a centralized communications activity to help all consumers understand how to use the meters and obtain the benefits. The roll-out began in 2014 (DECC, 2012c).

The Energy Act was amended in 2011 to introduce the Green Deal and the Energy Company Obligation. The Green Deal was expected to provide the opportunity for households and businesses to improve their energy efficiency at no upfront cost. Paying back through future savings in energy bills and the Energy Company Obligation (DECC, 2012d) allowed supplier subsidy and integration with the Green Deal Finance. The Green Deal Finance Company (GDFC) was established by the government as a non-profit entity to support the Green Deal scheme by providing finance to households and landlords, enabling them to improve the energy efficiency of their properties.

The Department of Energy and Climate Change (DECC) proposed £3 million to be invested to train Green Deal Assessors and to address skill shortages (DECC, 2012b). Estimates suggested as many as 65,000 jobs could be created by the Green Deal, with up to 54,000 of these being for installers (DECC, 2012b). Gleeson et al. (2011) contend that there was the need for 'a substantial increase in the capacity of the construction industry to carry out this work and training programs to produce the necessary skills' (p. 3) to meet the demand for skilled workers. For the longer term, the DECC declared it would be working closely with the governments advisory bodies on sector skill needs, the Sector Skill Councils, employers and the Department for Business, Industry and Skills (BIS) to develop new apprentice schemes to support the demand likely to be generated by the Green Deal (DECC, 2012d).

While the Green Deal was a well-resourced programme, research within the UK and across Europe (Gleeson et al., 2011; CEDEFOP, 2010) pointed to the additional demands on the labour market that should also be a factor in its preparation. To not do so left open the risk that labour may not be available to do the work in the numbers and with the skills required. As it transpired, funding for the Green Deal was discontinued in 2015 and the potential for a skills shortage disappeared. The Green Deal scheme had failed to sufficiently engage with householders and take up was much less than budgeted. The Department's main target of providing energy saving measures ahead of the March 2015 deadline was passed and, in July 2015, the Department announced it would not provide any further funding for Green Deal loans and the Energy Company Obligation (ECO) will end in March 2017, effectively bringing the schemes to a halt. In their April 2016 findings, the National Audit Office reported that while the Green Deal scheme provided energy saving in more than 1,000,000 homes ahead of schedule, they did not achieve the targeted CO_2 reductions which was the primary aim, the schemes saved substantially less CO_2 than previous schemes, to that date no more than 1 per cent of measures had blended finance from the Green Deal and the schemes had not improved the energy efficiency of many harder to treat homes as the Department initially expected.

The finance company established to manage the Green Deal loans, the Green Deal Finance Company, incurred substantial losses from the start as a result of the low demand for Green Deal loans. The company had been expected to become self-financing once revenue from fees and repayments was sufficient to cover operating costs, when loans were over £450 million. However, the finance company's loan book was just £17 million at the end of 2014, compared with the Department's impact assessment prediction of £695 million (UK NAO, 2016).

The government has yet to decide how it intends to proceed, decisions that will need to be made soon if emission reduction targets and carbon budgets are to be met. A lack of continuity in government energy-efficiency policies is likely to increase costs, as businesses require a higher return on risky investment in training, accreditation and capacity.

STAKEHOLDERS: INDUSTRY, TRADE UNIONS AND CIVIL SOCIETY

The EU formed the European Economic and Social Committee (EESC) in order to allow economic and social interest groups to express their views at a European level and to strengthen the 'democratic legitimacy and effectiveness of the European Union by enabling civil society organizations from the Member States' (Hache, 2011, p. 8). It subsequently refined the process by formally identifying sectors within civil society and recognizing peak representative organizations including recognized employers' organizations and trade unions. However, and unlike other countries across

Europe and the European Union, the UK does not afford statutory recognition to the peak representative organizations of business and workers.

In the UK, the peak representative organizations for employers and workers are the Confederation of British Industry (CBI) and the Trades Union Congress (TUC).

Business/Employers' Organizations

Historically, industry representation in the UK has been largely sectoral, whereby associations of business with similar interests have formed to undertake the representative and collective role on all matters related to their sphere of interest. For most activities these associations do not collaborate and advocate collectively. As observed by Bailey and Rupp (2006, p. 44) 'lobbying has historically been their [UK associations'] main activity but, due to a profusion of associations – often in competition with each other – and lack of a strong peak organization, they have frequently been by-passed by the British governments'. They note that historically the CBI has often acted as a competitor to sector associations rather than as a peak organization, although with the introduction of the Climate Change Agreements (CCAs[3]), informal partnerships emerged between the CBI and the trade associations to press their case.

The CBI is the largest national business representative association in the UK. It is the UK affiliate on the European representative organization of business and employers, BusinessEurope, and for other international agencies such as the ILO and the OECD. In the UK and internationally, the CBI assumes the mantle of a peak membership and representative organization for business and industry associations, although it has neither the constitutional authority nor the charter to represent industry in its entirety. This has the effect of mitigating the impact of its advocacy and it can also be conflicted between the broader interests of industry in the UK and the interests of its corporate members.

The CBI has expressed its support for the ambitious targets of the UK government and contends that business is tackling the low-carbon challenge (Cridland, 2011). It says that business recognizes that unabated climate change could pose an unacceptable risk to the stability of ecosystems and the economy and that 'the businesses which will succeed in the twenty-first century will be those that seize the opportunity to adapt to a low-carbon future' (CBI, 2009, p. 4). The CBI identifies the role of business in the shift to a low carbon economy as to find and implement solutions, to provide consumers with information to make informed choices, to develop and deploy low carbon technologies, and to be an investor and innovator to ensure the UK can exploit its opportunities (CBI, 2012a).

In 2007, the CBI moved to the climate change 'moral high ground' when it formed its Climate Change Board and published its climate manifesto in the report *Climate change: Everyone's business*, written by the 17 business leaders who were the members of the Board (CBI, 2007). This was a declaration of the CBI commitment to climate change mitigation action and its belief that there was advantage in the UK being a leader. The report has, in the absence of a formally stated policy, become the CBI's manifesto on climate change. The report arguably was an influential reference in the public policy (Eurofound, 2011d). The key messages from the report are the government's targets for 2050 will stretch resources but are achievable and at a manageable cost – provided early action is taken; in the run up to 2020, the emphasis must be on much higher energy efficiency together with preparations for a major shift to low carbon energy sources; the UK has a unique opportunity to prosper in key markets of the future by taking a lead in the development of low carbon technologies and services in power, buildings, transport and industry; empowering consumers to make low carbon choices is equally vital; and market forces will drive big changes, but they will not by themselves be enough to do the job.

A distillation of the CBI's climate change activities finds their advocacy has focused on public policy and programmes, international activities and energy policy. The CBI's directives lack penetration into industry and economic policy and the subordinate issues of transport and agriculture and its advocacy is often about issues that are one-off and ad hoc rather than as part of a considered long-term plan to strategically influence policy.

While the CBI has been prolific in its written contributions to public debate (for example, the report *Climate change: Everyone's business* (CBI, 2007); the report *All together now: A common business approach for greenhouse gas emissions reporting* (CBI, 2009); and *Climate change and business: The role of business* (CBI, 2012a), it has not distilled its views into a clear policy statement. At one end of the scale it articulates a broad overarching agenda and the leadership role of business in the shift to a low carbon economy, while at the other end it talks in detail about the role of government to protect business such as the benefit of regulating the requirement to display energy certificates (CBI, 2011). However, it does not turn its attention to the process of transitioning and while the CBI in its advocacy demands policy coherence and dialogue across portfolios of government (CBI, 2009), there is no evidence that it has prepared advice about what it expects the government to do. The CBI's broader strategy also is unclear. Its various statements often relate to individual situations or specific programmes, critiques of proposals by the government that are the subject of the day, for example, the 'Heat is on' publication which is a response to the UK government's proposed policy for community heating (Cridland, 2012). The CBI's statements can also be contradictory, for instance with calls for the removal of subsidies paid by other governments within the EU to ensure a level playing field for UK business across Europe but it then calls on the UK government for subsidies to protect disadvantaged sectors (Cridland, 2011).

In 2014, the CBI revamped the Climate Change Board, creating the Energy and Climate Change Board as an advisory committee that describes itself as a group of business leaders committed to tackling the UK's triple challenges of energy security, affordability and de-carbonization. It was particularly active in the lead up to the 2015 Paris COP, releasing the report *Setting the bar: Energy and climate change priorities for the government* (CBI, 2015a), which offered advice on the business requirements of UK, EU and international policy.

While the CBI is mandated to advocate on behalf of its direct corporate members, its industry association members advocate on their own behalf except on issue-specific occasions when they agree that CBI should be the advocate. On climate change, the CBI is not briefed to act on their behalf. Bailey and Rupp (2006) provide a pithy assessment of business representation in the UK, observing that organizations experienced poor credibility with government because there were so many of them and they lacked a strong peak organization. As a consequence, the UK government often chose to deal directly with the leading companies. In that regard, the CBI is strategically positioned through the membership of its Energy and Climate Change Board, which is made up of the leaders of industry and the business contacts government requires to support and facilitate policy implementation.

The CBI is not an 'employer's organization' in the terms defined by Blackwell (1999a) as a collective organization of employers. Its constitutional mandate does not extend to representing its members in wage negotiations or making representations about such matters with government or trade unions. The labour market and the impact of climate policy on people and the workplace have not been identified by the CBI as issues that require special consideration. On the occasions labour market issues are mentioned in CBI communications, such as its 2011 report *Mapping the route to growth, rebalancing employment* (CBI, 2011), the issue on which it seeks action is education and training, advocating for a system that provides skills, capabilities and attitude for the workplace. It does not relate these reforms to the labour market reforms identified by others as necessary for the effective transition to a low carbon workplace (for example, ILO, 2012; OECD, 2011; Eurofound, 2011d).

A thorough search of government and media websites found little evidence that the public policy process took heed of the views and activities of the CBI, or more broadly business and industry associations. It does not appear that CBI exerts influence over public policy on climate change or that government, commentators or researchers look to CBI to inform their enquiries.

Trade Unions

The Trades Union Congress (TUC) is a peak UK trade union organization. All trade unions in the UK are affiliated with the TUC. The UK trade union movement is much more structurally rational than CBI and therefore accountable, with a collegiate structure and membership representation from across the breadth of industry.[4] Policy and decisions made by the TUC are observed and implemented by all member unions under a top-down governance model. A problem confronting the TUC is that trade union membership is declining, in both the public and private sectors.[5]

The TUC is the principal civil society advocate in the UK within its organizational mandate for labour relations and workplace issues. It is the nature of the UK system that these are matters over which the UK government has traditionally not been directly involved; instead, they are left to the parties directly involved. The statutory framework governing the labour market, to the extent it exists, is through international labour conventions, EU law and UK common law precedent. Wage agreements, while binding on the parties to the agreement, are negotiated at the workplace or, where the employer and the workers agree, an industry or sectoral agreement.

On climate change, the TUC defers to the policies of the International Trade Union Confederation (ITUC) and European Trade Union Confederation (ETUC). The trade union movement's research finds that the shift to a low carbon economy will be at the expense of some jobs, new jobs will be created and others will morph into new occupational profiles and skill sets (ETUC, 2007). Its guiding principles for the shift to a low carbon workplace are that it should deliver decent work and good jobs, and that during the transition the treatment of workers is fair and just (ITUC, 2012a). The UNFCCC climate agreements now also incorporate this social value. The agreements from the 2010 Copenhagen COP introduced a provision that signatories will, in considering the social dimension of climate change, ensure that workers are provided decent work, good jobs and a just transition.

The principles that underscore TUC's climate change advocacy are that the UK government must commit to delivery on environmental policy; the state has a central role in stimulating the green economy; the state must encourage research and development; and the state must ensure the workforce has the necessary skills (TUC, 2009).

The TUC's Green Workplaces project (TUC, 2010) opportunistically sought to favourably reposition the union during the transition to a low carbon workplace and promoting the central role of the trade union (Syndex, 2011). The project and related programmes are also being used by the TUC to leverage authority in workplace negotiation. The unions' Greener Deal Guide provided union workplace representatives an explanation of the climate change issues and also explained to members how climate change can be used as a vehicle for union renewal, how it can extend the union consultation agenda and how union involvement in the environmental agenda can bring new members (TUC, 2012a). The TUC was also campaigning for the right for trade union workplace environmental representatives to be permitted reasonable time off during working hours to promote sustainable workplace practices, carry out environmental risk assessments and audits, consult with management and undertake training (TUC, 2015).

The Green Workplaces project (TUC, 2010) is the TUC's flagship climate initiative and has been widely applauded across the EU and member states as a best practice model (Eurofound, 2011d). The TUC asserts that 'half of UK carbon emissions are produced by work activity . . . they are an obvious place to focus action on climate change as organizations are better placed than individuals to install cost-effective green measures' (TUC, 2012a, p. 1) adding:

> Green Workplace projects are workplace-based initiatives that bring together the practical engagement of both workers and management to secure energy savings and reduce the environmental impact of the workplace. (TUC, 2010, p. 4)

They do this through activities such as awareness-raising events, staff surveys and training workshops. The process is decided by the workplace actors but the objective is to achieve joint management and union environment committees and framework agreements that embed workforce engagement in decisions about the way organizations do their work (TUC, 2010).

While Eurofound and the TUC describe Green Workplaces as employer/trade union initiatives (Syndex, 2011), the employers involved are usually public sector and institutional employers. This could be a reflection of the unions' diminished influence in the private sector where union penetration has declined to 14 per cent. Private sector management may also be resistant because of the clearly stated objective of ensuring union involvement in management decision-making and their use as negotiating tools on workplace matters.

The TUC currently offers training and support in the climate change workplace projects. The training has been supported by funding from the

Department for Business Innovation and Skills (BIS) Union Modernisation Fund (TUC, 2010), which was intended to facilitate transformational change in the organizational efficiency or effectiveness of unions. A consortium of four unions has successfully bid for government to fund its Climate Solidarity project (TUC, 2010), a programme which aimed to build community/workplace links for greening working and living but was subsequently closed down prematurely by the Conservative–Liberal Democrat coalition government in 2010.

The effectiveness of the TUC in its advocacy on climate change and the environment is not easy to measure as there have been no demonstrable outcomes. While its Green Workplaces project is an excellent concept, its take-up is ad hoc and largely amongst organized workplaces and it does not appear to have either spread the word about climate change effectively or spread the influence of the union within the UK. An alternative approach less based on the traditional employer/union adversarial negotiating style that is apparent in the TUC advocacy and programmes could potentially open opportunities to extend the Green Workplaces to the private sector.

Civil Society

Civil society has been prominent in shaping UK public policy on the environment and climate change. However, it was not until the introduction of the EU ETS in 2005, which was a tool intended by the government to influence the behaviour of market actors, that civil society generally acknowledged the impact of climate change on economic activity.

Voluntary initiatives such as the Carbon Disclosure Project in 2000, a corporate social responsibility initiative of investment fund managers that sought information from companies on their greenhouse gas emissions, were successful in creating awareness but were not successful in driving meaningful change. Corporate social responsibility and socially responsible investing advocated by civil society NGOs were concepts taken up across industry and investors but fell short of achieving the change required to meet the challenge of climate change (Pfeifer and Sullivan, 2008).

A number of NGOs have completed programmes aimed to investigate the UK's energy efficiency. The World Wide Fund for Nature (WWF) initiative targeted energy companies in the UK. The WWF published research that showed how taking action to manage emissions was more cost effective than inaction, demonstrating that it is less costly to set a price on carbon through market mechanisms and to source alternatives to fossil fuels than to do nothing (WWF, 2003).

The Friends of the Earth's Big Ask campaign, a parliamentary petition (Early Day Motion 178) calling for a new law requiring annual cuts in carbon dioxide emissions of 3 per cent, is credited with influencing the UK government's decision to introduce climate change legislation and to commit to the target of an 80 per cent reduction in emissions (Hall and Taplin, 2007). The Big Ask campaign was launched in May 2005 with the drafting of a climate change bill, which the Friends of Earth then successfully convinced a cross-party group of Members of Parliament to send to parliament. The then Opposition, headed by David Cameron, added its support to the campaign and called for the Bill to be included in the next Queen's Speech to parliament. The campaign realized its objective when the Bill passed into law on 28 October 2008 (FoE, 2008).

Despite the legislative impact of NGO environmental activism through campaigns like the Big Ask, academic researchers have published that there is little evidence of a connection with the public and the consumer, who have been neither consulted nor invited to comment on the objective or the intended action by those NGOs or government. Further, members of the public mostly knew little about environmental issues and had little consciousness of energy saving or of the environmental taxes and levies they were paying (Dresner et al., 2006; Pfeifer and Sullivan, 2008).

CONCLUSION

This is the second of the two detailed case studies designed as part of this research to better understand the role of employers' organizations and trade unions in the development of climate change policy conducted through the lens of ecological modernization. Figure 6.1 provides a comparison of the policy and process of the UK government within an EM framework against the theory, the research contention and the EU. It demonstrates that the policy and process in place in the UK is only partially consistent with the EM framework. If it is accepted that the EM framework has strong relevance to environmental policy in industrialized countries, it is not possible to reasonably predict that the present UK policy and process will deliver on its objectives.

Evaluation of the UK's policy and strategies finds they are responsible in relation to the UNFCCC and EU commitments but, as the Committee on Climate Change has warned, further action is required to maintain the rate of delivery against the commitments (UK CCC, 2016). In 2014, the Committee reported to parliament that step changes in the pace of emission reduction are needed and the second and third national carbon budgets should be tightened, a recommendation on which the government declined to take action. In another assessment addressing adaptation measures, the Committee observed that some sectors are near the limits of their capacity to adapt further and that climate risks are not being fully incorporated into some major strategic decisions. An OECD (2010) review found that more could be done to align economic and environmental objectives that would unleash opportunities for green investment. It encouraged the UK government to progress the integration of programmes across portfolios to minimize the potential for conflicting regulation and to undertake labour market planning to ensure the workforce is available in the numbers and with the skills required for the new low carbon economy. There is also evidence in the UK of resistance to adoption of the policies amongst the major utilities and residential consumers (UK CCC, 2016, UK NAO, 2016). Investigations have also found that the UK's water companies have still not made any specific investment in climate adaptation to tackle potential shortfalls in water supply (DECC, 2011f).

The data establishes the impact of climate change policy over the labour market; however, the targets from the country's climate change and energy policies have been achieved without the effective interventions of the employers' organizations and trade unions and without an effective labour market plan. That said, both employers' organizations and trade unions are active and effective advocates within their respective peak European organizations, which the review of the literature for this research finds are an effective influence within the European Commission and over European policy.

NOTES

1. The Scheme enabled participants in the Scheme to buy and sell allowances to emit greenhouse gases. It was launched in March 2002, when the Department for Environment, Food and Rural Affairs (DEFRA) agreed to pay participants £215 million over five years in return for binding commitments by the participants to cut their greenhouse gas emissions. Since then, these participants and others have been able to trade emissions allowances. In the Scheme's first year (2002), the 31 participants receiving incentive payments reported total emissions reductions almost six times their total target for the year: 4.64 million tonnes, compared to a target of 0.79 million tonnes (UK NAO, 2016).
2. The 2016 referendum vote for Britain to leave the EU places further doubt on their commitment to the EU's climate change authority.
3. Climate change agreements were voluntary agreements made by UK industry and the Environment Agency to reduce energy use and carbon dioxide (CO_2) emissions. In return, operators receive a discount on the Climate Change Levy (CCL), a tax added to electricity and fuel bills.
4. The TUC says it is 'the voice of Britain at work. With 58 affiliated unions representing 6.2 million working people from all walks of life, we campaign for a fair deal at work and for social justice at home and abroad. We negotiate in Europe and at home build links with political parties, business, local communities and wider society' (TUC, 2012b, p. 1).
5. Private sector: 2005, 16.9%; 2010 14.2%; 2015 13.9%. Public sector: 2005, 58.2%; 2010, 56.4%; 2015, 54.8% (BIS, 2016).

7. Comparative analysis: country profiles and case studies

COUNTRY PROFILES

Each of the countries profiled submitted its voluntary commitment (INDC) to reduce GHG emissions and initiatives to pursue the objectives of the Convention in advance of the UNFCCC 2015 Paris COP.[1] The summary of relevant features of their INDC is at Table 7.1.

DETAILED COUNTRY PROFILES

The data for this qualitative analysis was gathered from a literature search of the climate change policy and the policy development process in the UNFCCC, EU and eight countries. The countries were selected as three of the industrialized and developed economies in the EU and five from other regions linked by their common origins with the British Commonwealth and the consequent influence over their legal and parliamentary systems, language and, to an extent, their cultures. China was not included.

Consistency in the data collection process was guided by a review of the regional and country economic and labour market situations, climate change exposure, climate change legislation and regulation and the peak employers' organizations and trade unions. A template was developed to identify the main activities of the employers' organizations and trade unions and their engagement with climate change as a service to their members and as an influence over the policy development process of their region and the state. The template is attached as an Appendix. Each of the organizations profiled were advised of the data search and were invited to contribute and, while some acknowledged receipt of the invitation, none of the organizations submitted any information or comment.

Table 7.2 is an overview of the key economic and labour force data for each of the eight countries profiled.

Table 7.1 Summary of the relevant features of the profiled countries' INDC

	National circumstance	Contribution	Contribution mitigation	Contribution adaptation	Fairness	Means to implement
Kenya	80% arid 75% GHG from land use	Reduce GHG emissions 30% by 2030 relative to BAU* subject to international assistance	Expand geothermal, solar and wind Tree cover of 10% Low carbon transport Climate smart agriculture	Mainstream adaptation into development plans subject to international finance and technology assistance	Poverty alleviation and economic development must be considered	US$40 billion by 2030 from domestic and international sources
Australia		Reduce GHG emissions by 26–28% below 2005 by 2030	The Emission Reduction Fund supports businesses to reduce emissions, supported by the Renewable Energy Target	To develop a National Resilience and Adaptation Strategy	The target doubles the rate of emission reduction, is a significant increase beyond the 2020 target, and is comparable with other advanced economies	Market mechanisms through the Emission Reduction Fund

Singapore	Urban density, limited land mass, low flat land, low wind speeds, lack of geo thermal resources	Reduce emission intensity (per S$ of GDP) by 36% from 2005 levels by 2030, and stabilize emissions aiming to peak by 2030	Limited option given the natural circumstance, very energy efficient	Programmes in place to address food security, infrastructure resilience, public health, flood risks, water security and coastline	Ambitious target given the limited options for action	Domestic sources
Canada	Growing population, extreme temperatures, large landmass and significant natural resources	Reduce GHG emissions by 30% below 2005 levels by 2030	Has established stringent coal fired electricity standards Stringent GHG emission standards for the transport sector			The government has in place legislative instruments coupled with significant investments in clean energy technologies
India	17.5% of world's population, 300 million live in poverty, 300 million without electricity and safe drinking water	Reduce emission intensity of GDP by 33% by 2030 from 2005 level	40% of electric power from non-fossil fuel sources by 2030 Create carbon sink of 2.5–3 million tonnes by 2030	Resilient agriculture Conservation of water Coastal zone management Disaster management	Considered fair and ambitious given develop-mental challenges	Ambitious global agreement International assistance requirements for Adaptation US$206 billion to 2030, $7.7 billion for disaster management, $834 billion for mitigation to 2030

Table 7.1 (continued)

	National circumstance	Contribution	Contribution mitigation	Contribution adaptation	Fairness	Means to implement
UK	Refer to EU					
France	Refer to EU					
Germany	Refer to EU					
EU	The EU and its 28 member states	Binding target of at least 40% reduction in GHG emissions by 2030 compared to 1990	Domestic legally binding legislation in place for the 2020 climate and energy package		The target represents a significant shift beyond current undertaking of 20%reduction by 2020 and is in line with the EU objective to reduce emissions by 80–95% by 2050 cf 1990	

Note: *BAU scenario of 143MtCO_2.

Table 7.2 Chapter profiles (UN, 2016)

	Kenya	Australia	Singapore	Canada	India	UK	France	Germany
Region	East Africa	Oceania	SE Asia	Nth America	Southern Asia	Northern Europe	Western Europe	Western Europe
Population	45.5 million	23.6 million	5.5 million	35.5 million	1.27 billion	63.5 million	64.6 million	82.6 million
Urban population	25.5%	89.3%	100%	81.7%	32.4%	82.3%	79.3%	75.1%
Capital city population	3.7m	Canberra 415,000	Singapore 5.5 million	Ottawa 1.3 million	New Delhi 250,000	London 10.2 million	Paris 10.7 million	Berlin 3.5 million
GDP (US$m)	54,443	1,531,282	295,744	1,838,964	1,937,797	2,678,455	2,806,432	3,730,261
GDP per capita (US$)	1,227	65,600	54,648	52,270	1,548	42,423	42,338	45,091
Labour force participation female	62% (N/A)*	58.8%	58.8%	61.6%	27% (89.9%)*	55.7%	50.7%	53.6%
Labour force participation male	72.4% (N/A)*	71.8%	77.2%	71.0%	79.9% (83.1%)*	68.7%	61.6%	66.4%
Unemployment	9.2%	5.7%	2.8%	7.1%	3.6%	7.5%	10.4%	5.3%
Population 0–14 years	42%	19.1%	15.7%	16.5%	28.8%	17.6%	18.2%	13.0%

Note: *Informal economy labour market participation.

INTERNATIONAL ORGANIZATIONS

As is seen across the organizations profiled in this study, many of the national organizations rely on the policy papers and research of the peak international organizations to guide their domestic advocacy on climate change and as such frames the civil society contribution to the domestic policy and legislative process.

Labour market planning and the employment and workplace impacts of climate change have not been a significant issue for these organizations. The International Chamber of Commerce (ICC) and the International Organisation of Employers (IOE) have not commented on the labour market impacts of climate change, the ICC has not engaged in the green jobs activities of the International Labour Organization (ILO), and even the International Trade Union Confederation (ITUC) advocacy addresses only its decent work and just transition mantra. The Green Jobs campaigns initiated by the ILO and supported by UNEP, IOE and ITUC go some way further but maintain the relatively narrow focus of jobs that are considered green, potentially creating a demarcation between the green and not green jobs.

The ICC is the principle vehicle for business and industry stakeholders to interface with the negotiations for the international agreements generated through the UN system. It is the UN-nominated coordinator of the Business and Industry Major Group of the nine major groups through which stakeholders are coordinated. That said, it has not been given a mandate to represent all those stakeholders and so serves the dual role of secretariat on behalf of all stakeholders and advocate for the Chamber of Commerce movement.

ICC activities are guided by its Environment and Energy Commission, which is a committee comprised of interested members from national Chambers of Commerce and corporate members. The policy documents prepared by the Commission during 2015 framed the ICC advocacy in the lead up to the Paris COP and their desired outcomes from the negotiations and the content of the Agreement.

Important to note is that the ICC does not seek to represent the interests of business as employers, rather its mandate is 'to promote international trade and investment through a rules-based multilateral system' (ICC, 2015b).

The IOE mandate is in respect of the interests of business as employers and with a particular focus on the employer interface with the ILO. It does not engage in either the ICC Environment and Energy Commission or the Business and Industry NGO (BINGO) as a stakeholder at the UNFCCC negotiations. The IOE is not an active advocate on climate change and when required to comment it defers to the formal position adopted by the ILO which supports the principles that underpin a just transition and decent work (ILO, 2012). The IOE involvement with climate change has been limited to ensuring there is balance in ILO commentary and in matters such as the Green Jobs Report, the UNEP Green Economy research project and report, and to be able to advise members when requested. Many IOE members are also representative of the broader business interests of members, not only as employers.

It is also to be noted that the IOE mandate is quite narrow and its members retain the right to advocate on their own behalf. This leaves the ITUC to lead the debate in many areas and much more broadly than the IOE and accordingly is often unchallenged.

The ITUC is a vocal campaigner, albeit on the broader social agenda and the narrow just transition platform. The ITUC and its predecessor organizations have attended all climate change conferences of the UNFCCC and are conveners of the Trade Union NGO (TUNGO). As distinct from the ICC the ITUC has a mandate to advocate on behalf of trade unions in the UN system and elsewhere. It is also very well networked with other CSOs of a like mind.

The discussion about whether the ITUC is effective in the climate negotiations requires considerable analysis of a range of factors that go beyond the scope of this book. Prima facie the ITUC is professional in its approach and has made climate change part of its work programme. However, its interest is clearly the achievement of its social agenda and the just transition of workers which it advocates can only be delivered by the effective management of climate change through mitigation and adaptation and through climate change policy that delivers a just transition and decent work.

In the EM context, these organizations have not advocated that employment and the labour market are issues that require special treatment beyond the ITUC advocacy for a just transition and decent work which some would contend is an equity rather than labour market demand. However, as civil society the ICC and ITUC are formidable constituents and recognized as such not only for their informed advocacy but for their extensive reach across their constituencies that are integral to the effective implementation of policy and influence in their domestic jurisdictions.

EU AND THREE MEMBER COUNTRIES WITH INDUSTRIALIZED AND DEVELOPED ECONOMIES

The EU in its INDC noted the IPCC advice that action by developed countries is necessary and they must, as a group reduce their emissions by 80–95 per cent by 2050 compared to 1990 and their strategies intended to achieve this objective must be reported. It is also the only INDC among the countries profiled to extend their proposed emission reduction projections beyond the 2030 target. The EU committed targets were in each case a material extension beyond present achievements and were supported by well-structured action plans.

The three countries profiled from the European Union group are highly industrialized and mature economies. Their strategies are based on the three pillars of reducing emissions by increasing energy consumed from renewables, reducing primary energy use and energy efficiency.

COUNTRIES FROM OTHER REGIONS

The countries profiled for this section are Australia, Canada, Singapore, India and Kenya. Principal among the mitigation strategies of these countries were the shift away from fossil fuels towards renewables, energy efficiency and the role to be played by developments in technology. Adaptation measures were most advanced in the climate-exposed regions, which drew attention to their exposure to extreme weather patterns, extremes of heat and cold and coastline management. The individual strategic approach that underpins the Intended Nationally Determined Contributions (INDCs) submitted to the UNFCCC in 2015 by each of these countries can be summarized as:

- The Australian government has simply agreed to a target that is a mid-point of the targets of industrialized countries.
- Canada is proposing stringent emission standards for the electricity and transport sectors.
- Singapore, which has the least capacity to reduce resource consumption and introduce energy-efficiency measures, has still set an ambitious target, committing to research and development to advance technology.
- India and Kenya are focusing on wind, solar and carbon sinks, essentially reducing the growth in emissions while addressing the extension of the energy accessibility across the population. However, each has qualified their commitment as being contingent on the

> achievement of an ambitious global agreement in Paris, with the requirement for material financial and technology support from developed countries, and with consideration of the overriding priorities of poverty alleviation and the dependence on economic development to support social programmes. The cost of India's proposed climate change initiatives must almost entirely be borne by the rest of the world if they are to be implemented.

The INDCs are in the main fair and ambitious. Australia has been criticized for doing less than it is capable of, the contributions of both India and Kenya are conditional on financial and technology assistance and Canada's pledge will require new rather than scaled-up programmes. Only the EU extends the boundaries. While the aggregate of the INDCs of all Parties to the UNFCCC climate change agreements falls short by a significant margin of achieving the objectives of the Convention and the Paris Agreement, the INDCs from the profiled countries demonstrate that climate change mitigation and adaptation strategies are embedded in domestic policy and programmes, it is only the capacity to deliver that is the variable.

EMPLOYERS' ORGANIZATIONS AND TRADE UNIONS

The effectiveness of the peak national employers' organizations and trade unions in the profiled countries was analysed.

The regional and international organizations show an inconsistent interest in labour market issues. On the trade union side, the International Trade Union Congress and the European Trade Union Congress are active and strong advocates on the labour market platform and decent work and a just transition. On the other hand, BusinessEurope is a stronger advocate of the broader business interests as is the International Chamber of Commerce, while the International Organisation of Employers has a passive interest with interventions only as are demanded and in relation to their engagement with the International Labour Organization.

The national employers' organizations and trade unions are not as a matter of course expressly engaged in the domestic climate change policy process. The organizations in Australia, Kenya, Singapore, India and Germany have no dedicated climate change or sustainability activity, while the Canadian, French and UK organizations are engaged. It is only the French organizations MEDEF, UK Trades Union Congress and the Canadian Labour Congress that advocate on the labour market impacts.

The Chambers of Commerce in Australia, Canada, Germany and India, the UK Confederation of British Industry and MEDEF are active contributors to the climate change policy process on the business issues.

The effectiveness of advocacy in some of the countries is compromised because there are multiple peak organizations. In two of the countries, France and India, there are multiple peak organizations representing their members to government and at the regional and international organizations, and in Canada there is no peak employers' organization.

In France, while MEDEF is the peak employers' organization, representing over 750,000 companies of all sizes throughout the country among 76-member trade federations, the CGT is acknowledged as the largest of the five peak trade union confederations and in terms of electoral votes is the second largest by member numbers.

India has a proliferation of peak business representative organizations embracing the sectoral and regional groups. The research for this book examined two of the four employers' organizations that have input to the International Employers Organization. The Federation of Indian Chambers of Commerce and Industry is the only organization affiliated with the International Chamber of Commerce. Goyal and Sharma in their 2012 critique of the engagement with the government by the peak business organizations observe that they are of little interest to, nor are they influencers of government; rather, influence in policy rests with the CEOs of the major members' companies who are invited by government to consult. The three peak Indian trade unions profiled in this study are very similar in their objectives and structure and, while membership numbers are large, they are still a small percentage of the workforce. They are also generally known to be the trade union branch of a political party, as the AITUC is affiliated with the Communist Party and the INTUC with the Indian National Congress. The Confederation of Free Trade Unions of India boasts of its non-political affiliation and is emerging as a new force organizing those that are presently overlooked by trade unions and bringing together other unions as affiliates.

In Canada, the business associations, employers' organizations and trade unions in the provinces and regions hold great authority and often act separately from the peak national organizations and in representations to government. The vast distances and areas imposes regionally unique concerns that are often difficult to address within a national context and policy.

COMMENT

The profile of employment and workplace issues was raised in 2007 and 2008 in the context of climate change. The UNFCCC Bali Action Plan (UNFCCC, 2007) required the Parties to consider the economic and social consequences in their negotiations that resulted in the 2009 draft agreement to the Copenhagen COP and in the outcome agreements from the 2010 Cancun COP the requirement to provide workers decent work and a just transition. The ETUC (2007), UK government (GHK Consulting, 2007) and ILO (Worldwatch Institute, 2008) reports introduced the labour market issues that arise from climate change policy and introduced the green jobs concept and the need to support workers in the transition to a green economy. While the trade union movement had been an active participant in the international negotiations since the World Summit of 1992, it was not until 2009 that the International Organisation of Employers (IOE, 2009) considered whether climate change was an issue of concern for employers' organizations and released its climate change information paper. Despite this body of intellectual and practical evidence, Glynn (2014, p. 197), in his research into the role of employers' organizations and trade unions, concluded that:

> the beneficial contribution to the process by employers' organizations is conditional in theory and interest and capability. The contention holds in theory and ambition in respect of trade unions, but the practice finds that they are required to direct their resources to those activities of current interest and in which the union can deliver an outcome.

This finding is in respect of the organizations as domestic activists, the relevant statutory and regulatory jurisdiction, findings that are generally consistent in peak national organizations profiled here.

In ecological modernization terms, the country profiles mostly demonstrate optimum models of EM in policy and practice. In the developing countries, policies reflect some commitment to the application of optimum EM (notably Kenya) but the practice fails to implement the commitment, essentially due to the weak institutions of the state, the conditional nature of their commitments based on the contributions by other countries for technology and finance, that governments do not engage effectively with civil society or to influence ecological consciousness.

Table 7.3 rates the governments' commitment to an EM model. Interestingly, the Kenyan government policy represents a model of strong EM whereas the practice finds that the President has delayed the climate change legislation approved by government on the grounds that civil society was not engaged sufficiently in the process, and the ability to

Table 7.3 EM template overlaid on INDCs

	Kenya	Australia	Singapore	Canada	India	UK	France	Germany	European Union
State	10**	5	10	10	5**	10	10	10	10
Innovation and technology	10	5	10	10	5*	10	10	10	10
Industry	5	5	10	5	5	10	10	10	10
Civil Society	10	5	10	10	0	10	10	10	10
Ecological consciousness	10	5	10	5	5	10	10	10	10
	45	25	50	40	20	50	50	50	50

Notes:
* Government is actively pursuing the introduction of new technology and finance from other countries though the technology transfer arrangements under the UNFCCC agreements.
** Subject to international assistance.

implement the government's climate change plan is conditional on other countries' agreement to transfer technology and to provide the necessary finance. The recent change from a Conservative to Liberal Canadian government has seen a significant policy shift towards a model of strong EM but the Canadian government does not have the legislative authority over the provincial governments that are presently fossil fuel dependent economies. The Australian government, while declaring support for the international process and commitment to the objectives with reasonable targets for GHG emission reduction, is still considered to be only a moderate contributor to the global effort.

CONCLUSIONS FROM THE COUNTRY PROFILES

The EU and UK detailed studies provide a focus for comment about the issues that arose generally across all of the countries that were profiled.

Policy Framework

The EU's preparedness to embark on a costly programme of reducing GHG emissions when other major emitters would not make a similar commitment has already been discussed. Similarly, it is appropriate to question why the EU is prepared to maintain its faith in its social partners when in some areas of responsibility, they are underperforming. These concerns are related to two main factors. First, the continuing steady decline in membership of trade unions raises questions about their representativity. In 2013 trade union membership in France was 7.7 per cent of the workforce and Germany 18.1 per cent. In the UK, it was 25.1 per cent in 2014, which breaks down to 56 per cent of workers in the public sector and 14.2 per cent in the private sector (OECD, 2016). Second, in order to ensure good governance in environmental policy, a greater participatory approach is required than currently exists. There is a broad community of stakeholders in environmental policy development that includes industry, finance and commerce, employment, environment and consumer and citizen interests. Broader participation brings a higher degree of legitimacy to the process and allows the pooling of resources (van den Hove, 2000). The challenge lies in the practical design of the engagement process and ensuring the legitimacy and representativity of the interest groups. These are issues separate from those related to the social partners' concerns and are addressed through different regulatory instruments (van den Hove, 2000).

Until the Climate Change Act 2008, the UK's approach to the management of its climate change responsibilities was a blend of hard and soft policy, using tools such as the 2001 voluntary Climate Change Levy (CCL) and corporate social responsibility initiatives. While business in the UK responded positively, these initiatives were found to have little impact on consumer behaviour (Pfeifer and Sullivan, 2008).

The targets set in the EU 20:20:20 Climate and Energy Package Framework (Europa, 2010a), the UK Climate Change Act 2008 and the long-term commitment to reduce greenhouse gas emissions by at least 80 per cent by 2050 relative to 1990 levels formed the framework for the UK Government's emission reduction and energy-efficiency strategies (DECC, 2010).

Under the Kyoto Protocol, the UK committed to reduce emissions to 92 per cent of 1990 levels by 2012 (UNFCCC, 1997). Although this target was met by 2008, the question under consideration was whether the measures implemented could continue to deliver further emission reductions. To date, the UK has been able to harness its excessively inefficient use of energy across the economy. It also 'benefited' from the global financial crisis and the consequent reductions in economic activity, which it is argued was a contributing factor behind the large reduction in emissions in 2009 (DECC, 2011e). This is not to understate the UK government's achievements or the very positive work that has been done, but given that the UK has been coming off a low base of widespread inefficient use of energy, it remains to be seen whether it can continue the rate of progress.

Employment and Workplace Issues

The UK country study in the ILO report *Skills for green jobs: A global view synthesis report based on 21 country studies* (Strietska-Ilina et al., 2011) finds that there is no centralized national framework for the UK labour market and that, although the government has acknowledged skills gaps and shortages as a potential threat, 'these are through generalized statements rather than specific policy measures' (p. 421). The study concludes that the UK government's environmental strategies do not generally have a significant skills development component, iterating concerns that the labour required to meet the demands in the transition to a low carbon market may not be available in the numbers and with the skills required.

The reports by Gleeson et al. (2011), the ILO's (2011d) *Skills and occupational needs in green building* and the OECD's (2011) *Green growth strategy synthesis report* analyse the issues confronting the construction sector labour market as property owners move to reduce energy consumption and emissions generation. Gleeson's study of the UK finds that a substantial increase in the capacity of the construction industry labour market will be required to meet the demand created by a national retrofitting programme (Gleeson et al., 2011). Gleeson also suggests that the nature of the nation's green construction will change, creating new occupational profiles and occupations and requiring new sources of labour to meet the increased demand. These issues change the risk profile of policy and particularly the national Green Deal retrofitting initiative.

The EU's various charters cover a wide range of social responsibilities. They are comprehensive and are a material demonstration of the commitment by the member states to the social priority. The Community Charter of the Fundamental Social Rights of Workers established the major principles on which the European labour law model is based and shaped the development of the European Social Model. The EU adopted a Social Action Programme to implement the Charter, which was instrumental in launching initiatives in employment and industrial relations policy (Eurofound, 2011b). By contrast, the UK government would not vote in favour of the Charter, which meant that it could not be integrated into European law as a Treaty but could still be used as an interpretative guide in litigation concerned with social and labour rights. The Charter was subsequently incorporated as a Treaty in 2007 as the Charter of Fundamental Rights of the European Union, although it contains exclusions for the UK and Poland. The Charter breaks new ground by including a single list of fundamental rights which include not only traditional civil and political rights but also social and economic rights including the rights of freedom of assembly and freedom of association, of workers' representatives to information, to collective bargaining and of access to placement services, maternity and parental leave, social security and social assistance (Eurofound, 2011d).

Employers' organizations and trade unions are the organizations appointed to represent the interests of employers and the workforce. Under European regulation, employers' organizations and trade unions are afforded the title of social partners. In most EU member states, social partners are involved with low carbon economy issues from the stage of policy formulation, where they express their positions on policy proposals either through institutional tripartite bodies dealing with sustainable development or by direct lobbying on draft legislation (EC, 2011a). The EU views social partners as playing an important role in the economy as a whole and the labour market in particular. With respect to climate change, it looks to them to create consensus on the implementation of policies across industry and society. This model, however, does not flow through to the relationships with the state in the UK, where the peak organizations do not have a statutory role and are much less overtly involved with the development of domestic policy including climate change policy.

Business/Employers' Organizations

In the UK, the CBI's early work on climate change earned respect across industry and government for its contribution to public debate and guidance to members (Eurofound, 2009). However, since 2007 and its early flush of leadership, the CBI has retreated to providing only a passive programme that informs members through forums and newsletters and responding to government/public sector initiatives. The reactivation of the CBI advisory committee, the Energy and Climate Change Board, has reinvigorated its advocacy and policy direction and potentially its engagement with government.

In terms of addressing climate change, the CBI fulfils the role of providing the interface with industry but it is not directly instrumental in the development or implementation of climate change policy in relation to business and industry in the UK. This is a point of difference with BusinessEurope, which is an effective lobbyist on matters clearly within its mandate, which is well defined and understood across the EC. The European Trade Union Institute (ETUI, 2011) is less complimentary towards BusinessEurope, observing that it is unwilling to look at alternative forms of regulation if they can prevent recourse to legally binding acts. This observation is one that BusinessEurope would probably concur with. Climate change is a global challenge that requires global actions. Through its actions, BusinessEurope demonstrates that is committed to and aware of the challenges that climate change presents as well as the impacts of human activities.

Trade Unions

The union movement internationally has successfully advocated for the inclusion in global climate agreements provisions for a just transition and decent jobs, provisions that most governments have agreed to implement (UNFCCC, 2010). In the UK, the TUC advocates the implementation of these commitments and has also rolled out Green Workplaces as its major domestic initiative.

While responsible for a well-defined programme of advocacy, there is little to indicate that the TUC has any influence on the development of public policy or programmes in the UK. The TUC's policy priorities are not reflected directly in UK public policy. The unions' poor penetration of the private sector mitigates the impact of its advocacy with policymakers and underscores the criteria for selection of programmes to which it will commit resources; that is, programmes that create opportunity for it to re-engage with workers and to demonstrate its relevance (BIS, 2012).

By contrast, the ETUC is an active and committed campaigner for the social and environmental dimensions of climate change. The role of the state is central to its advocacy, contending the state should introduce binding regulation to address climate change supported by changes to relative pricing through taxation and investment in research and infrastructure (Laurent, 2010). The literature establishes that its affiliated unions are equally as diligent, although their approaches are nuanced by the prevailing domestic economic, social and political concerns and the extent of possible job losses, as in the case of Poland's unions in the coal mining and power sectors which were concerned that EU and ETUC GHG emission reduction proposals would adversely impact their employment opportunities (EC, 2011a).

Case Study Conclusions

UK employers' organizations and trade unions are well informed and their actions are supported by programmes with good research. Civil society environmental organizations, notably, have been and remain a major influence over public policy in the UK. However, since their early flush of activity, business and labour advocacy on public policy has become a response to proposals of government rather than advocating a strong policy position and they have not sought to retain a public profile on climate change policy. When the UK is compared to other countries in Europe, this is also the situation, and the research found that in practice, the contribution of the social partners and their commitment is dependent on their work programme which does not always prioritize climate change but rather reflects the current issues of the day (Eurofound, 2011b; CEDEFOP, 2010; OECD, 2013a).

In sum, the profiles that have been constructed provide a body of data that informs the analysis and from which it is found that industrialized countries have generally committed to optimum models of EM and the ecological modernization framework is an effective guide in the policy development process. Australia is an exception here, with a sub-optimum model. The value of the EM model in the situation of developing countries is still to be established. While the EM model can guide the policy development process, the agencies of government, industry and civil society lack the capacity to fulfil their roles under an EM framework. To that end, further research is required to establish how EM may be operationalized. At this stage, when the EM template is applied to India it appears that the present model is sub-optimal. Importantly, the engagement of national employers' organizations and trade unions in the climate change policy development process is dependent on the national circumstance and the priority afforded the issue relative to the other demands on their limited resources.

NOTE

1. Further to the negotiations under the Ad Hoc Working Group on the Durban Platform for Enhanced Action (ADP), the Conference of the Parties (COP), by its decision 1/CP.19, invited all Parties to initiate or intensify domestic preparations for their INDCs towards achieving the objective of the Convention as set out in its Article 2, without prejudice to the legal nature of the contributions, in the context of adopting a protocol, another legal instrument or an agreed outcome with legal force under the Convention applicable to all Parties. In decision 1/CP.20 it is further specified that in order to facilitate clarity, transparency and understanding, the information to be provided by Parties communicating their intended nationally determined contributions may include, as appropriate, inter alia, quantifiable information on the reference point (including, as appropriate, a base year), timeframes and/or periods for implementation, scope and coverage, planning processes, assumptions and methodological approaches including those for estimating and accounting for anthropogenic greenhouse gas emissions and, as appropriate, removals, and how the Party considers that its intended nationally determined contribution is fair and ambitious, in light of its national circumstances, and how it contributes towards achieving the objective of the Convention as set out in its Article 2 (UNFCCC, 2015).

8. Perspectives on the governance quality of climate policymaking

Having looked at the contribution that EM can play in evaluating the performance and outcomes of climate-related policy processes in a range of countries and case studies, this chapter focuses on the perspectives of participating stakeholders themselves. Survey respondents were identified using the Internet search terms 'employers' organization' or 'trade union' and 'participants' list', contacted by email and invited to participate in an anonymous online survey. The survey was in English only. Respondents were from a diverse range of continents, geographical regions, economic trade zones and developed and developing countries. These included Albania, Anguilla, Armenia, Australia, Bangladesh, Belgium, Canada, France, Georgia, Germany, Ghana, Guyana, Haiti, India, Jordan, Kenya, Latvia, Nepal, New Zealand, Norway, Pakistan, the Philippines, Portugal, Romania, Sweden, the United Kingdom and Ukraine. A number of respondents chose not to identify their country, indicated that it was 'not relevant', pointed out that they or their organizations were active in multiple countries or chose to identify themselves on a sub-national level (for example, Wales).

For analytical purposes, respondents were asked to categorize themselves as 'employers' organization', 'trade union' or 'other', identify their level of engagement in climate policy (intergovernmental, regional, national, multiple, or other), their region (EU or other) and their gender. 'Other' respondents included respondents who identified as 'civil society organizations other than employers' organizations and trade unions' (4), local government (2), other areas of activity, combined with trade union (3) and education (1). Those respondents who selected 'multiple levels' were from all sectors and active on all levels; consequently, for example, although no employers' associations were only active on the intergovernmental level, two were active at multiple levels, which included the intergovernmental level. 'Other' levels of policymaking identified were local (1), state – that is, sub-national government (2), individuals active within their specific sector (4) or individuals who indicated they had no formal role (4).

A list of survey respondents and their status is included in Table 8.1.

Respondents were further asked to provide a rating for their perceptions

Table 8.1 List of survey respondents

Participants	Employers' organization	Trade union	Other (please specify)	Total
Initial cohort	17	64	10	91
Level of activity				
Intergovernmental (e.g. UNFCCC)	0	2	0	2
Regional level (e.g. EU)	2	9	0	11
National level	5	32	6	43
Multiple levels	5	15	2	22
Other (please specify)	1	6	4	11
Region				
EU	5	23	3	31
Other	10	37	5	52
Gender				
Female	6	16	3	25
Male	10	40	4	54
I prefer not to respond	0	2	0	2
Total surveys commenced	*13*	*57*	*6*	*76*
Total surveys completed	*9*	*41*	*4*	*54*

of their role in climate policy negotiations, using a Likert scale of 1–5 ('very low' to 'very high') following the indicators of Table 8.1, thereby generating a score out of 55. Respondents were also invited to make comments underneath the appropriate indicator. Table 8.2 contains a summary of survey questions.

A number of caveats should be made here regarding the results presented below. The most obvious is that the results constitute the outcomes of a small-*n* survey only. Clearly the number of employers' organizations responding to and completing the survey is much lower than trade unions, and these results should be seen as largely anecdotal. This is even more so than those who identified as 'other'. In these instances, the numerical values attributed to these responses might appear to have more statistical authority than is actually the case. There are larger numbers of union respondents and therefore the results may be given a little more credence, but again, it must be stressed that given the global numbers of unions worldwide, the results are not broadly representative. The same may be said for the EU and non-EU results. It should be further added that given the size of the trade union cohort, these regional views are also largely influenced by those responses. On

Table 8.2 Summary of survey questions

Indicator	Question
Inclusiveness	*Do you consider the policy negotiations in which you participate are inclusive of your interests?*
Equality	*Do you consider the policy negotiations in which you participate treat everyone equally?*
Resources	*What level of resources do these policy negotiations provide for you to participate?*
Accountability	*Do you consider the policy negotiations in which you participate act in an accountable manner?*
Transparency	*Do you consider the policy negotiations in which you participate act in a transparent manner?*
Democracy	*Do you consider the policy negotiations in which you are involved act in a democratic manner?*
Agreement	*Do you consider the manner in which agreements are made in the policy negotiations in which you are involved is effective?*
Dispute settlement	*Do you consider the manner in which disputes are settled in the policy negotiations in which you are involved is effective?*
Behavioural change	*Do you consider the policy negotiations in which you are involved will change the behaviour that leads to climate change?*
Problem-solving	*Do you consider the policy negotiations in which you are involved will solve the problem of climate change?*
Durability	*Do you consider the policy negotiations in which you are involved will be durable?*

Note: Explanatory text and introductory materials omitted.

this view, the results should be seen as providing an indicative picture that only represents the aggregated perceptions of those who answered the survey.

RESULTS OF THE GOVERNANCE ANALYSIS

Table 8.3 Commentary on Results

Some interesting results emerge when all levels of climate policymaking are combined to produce an impression of respondents' perceptions regarding the governance quality of climate policymaking generally. Employers' organizations provided the highest total score (49 per cent; all percentiles are rounded to the nearest whole number). But this was a 'fail' as it did not pass the 50 per cent threshold. At the principle level,

Table 8.3 Sector perspectives on the quality of climate governance with all levels combined

Principle	1. Meaningful Participation Maximum score: 25 Minimum: 5							
Criterion	*1. Interest representation* *Maximum score: 15* *Minimum: 3*				*2. Organizational responsibility* *Maximum score: 10* *Minimum: 2*			Principle Score
Indicator	**Inclusiveness**	**Equality**	**Resources**	*Criterion Score*	**Accountability**	**Transparency**	*Criterion Score*	
Employers	2.98	2.67	**1.31**	***6.96***	**2.36**	**2.19**	***4.54***	**11.5**
Unions	2.52	**2.36**	**1.86**	***6.74***	2.54	2.59	*5.13*	**11.87**
Other	**1.42**	**1.38**	**1.25**	***4.05***	**1.9**	**1.69**	***3.59***	**7.64**

Principle	2. Productive deliberation Maximum score: 30 Minimum: 6								
Criterion	*3. Decision-making* Maximum score: 15 Minimum: 3				*4. Implementation* Maximum score: 15 Minimum: 3				Principle Score
Indicator	**Democracy**	**Agreement**	**Dispute settlement**	*Criterion Score*	**Behavioural change**	**Problem-solving**	**Durability**	*Criterion Score*	
Employers	3	2.59	2.5	*8.09*	2.73	**2.47**	**2.28**	***7.48***	15.57
Unions	2.74	**2.47**	**2.36**	*7.57*	2.55	**2.25**	2.62	***7.41***	**14.98**
Other	**1.38**	**1.35**	**1.5**	***4.23***	**1.75**	**1.75**	**1.75**	***5.25***	**9.48**
Total (out of 55)									
Employers									**27.07**
Unions									**26.85**
Other									**17.12**

Note: Light grey represents the highest score; dark grey the lowest; numbers in bold are below the threshold value of 50 per cent; average results from each sector.

the sector awarded deliberation the highest score overall but participation was a 'fail'. At the criterion level, interest representation, decision-making and implementation received the highest scores overall but organizational responsibility, interest representation and implementation were all 'fails' (although implementation was rated as the highest 'fail' among the three sectors). Democracy was the highest rated indicator overall (3), closely followed by inclusiveness (the second highest indicator), with resources the lowest (1.31). Several indicators did not meet the threshold: transparency (the second weakest indicator) and accountability, which explains the failure of organizational responsibility at the criterion level; and durability and problem-solving, which further explains the failure of implementation as a criterion. The performance of other indicators can only be characterized as 'average', with dispute settlement achieving only a borderline 'pass'. This was not a strong performance and the failure of key governance values such as organizational responsibility and implementation is not a good sign. The only positive intimation is that employers rated the inclusiveness of climate policymaking positively. Those employer organizations that took the survey, while small in number, did not appear confident that current policymaking venues would tackle climate change effectively and they had some serious concerns about transparency and accountability.

Trade unions provided the second highest total score – a 'fail' similar to that of employer organizations (also 49 per cent). Both principles were also a 'fail', although participation was the highest scoring 'fail' of the three sectors. Decision-making and organizational responsibility were both a 'pass', with organizational responsibility being the highest scoring criterion; interest representation and implementation did not meet the threshold. It should be noted that were it not for democracy, decision-making would otherwise have been a 'fail' at the criterion level. Resources was again the lowest rated indicator (1.86), while democracy was once again the highest (2.74). This sector also 'failed' several indicators, but these were greater in number and different from those of employers: problem-solving (the second lowest indicator), dispute settlement, agreement and equality. The remaining indicators were also 'average'. The results stand in contrast to employers, who gave a lower score to organizational responsibility than unions, who in turn gave a lower score to interest representation: in this regard it should be noted that in addition to the 'fail' for equality, inclusiveness only achieved a borderline 'pass'. This may reveal a difference between the two sectors regarding perceptions of access and influence in policymaking.

The lowest results by far were those from 'other' (31 per cent), but extreme care should be exercised in interpreting these, as the number of respondents was very small. These results are included to acknowledge those who chose to select 'other' as a category of respondent. Everything at the principle, criterion and indicator levels was a 'fail'. As with employers and unions, resources was the lowest indicator (1.25) and the lowest rated overall. Accountability was the highest but still a 'fail' (1.9).

Table 8.4 Commentary on Results

EU respondents provided the highest score for the governance quality of climate policymaking at the national level – the highest score to be achieved across the whole survey (57 per cent). Both principles and all criteria at the national level exceeded the threshold, with all receiving the highest scores. Resources was the lowest rated indicator and a 'fail' (2.32), while accountability was the highest rated indicator overall (3.16). All other indicators were a 'pass', mostly within band '3' range. This is a 'satisfactory' result, perhaps reflecting the fact that policymaking at this level could be perceived as having the greatest likelihood of successful implementation, with clear lines of electoral accountability and parliamentary scrutiny. Additionally, the EU has a history of interest intermediation through various corporatist-style approaches to institutional governance and policymaking (Kjaer, 2015), which may also have influenced perceptions.

The regional level was second (53 per cent), with both principles and three criteria meeting the threshold; interest representation was a 'fail'. Once again, resources was the lowest rated indicator and a 'fail' (2.06) and accountability the highest (3). Equality was a 'fail' as was inclusiveness (explaining the 'fail' for interest representation). Problem-solving was also a 'fail' and the second lowest indicator. It is tempting to read these results as a reflection on the EU itself as a forum for climate policymaking. If so it might be possible to identify a perception regarding asymmetries of power (perhaps between member states, as well as between sectors). Likewise, it may be that the ability of the EU to successfully reduce emissions also generated a negative attitude; here the problems with the EU ETS, such as the problem of the over-allocation of exemptions reducing demand and keeping the price low, leakage to non-ETS countries, and fraud spring to mind (Branger et al., 2015).

The intergovernmental level was third and a borderline 'pass' (50 per cent, but only as a consequence of rounding). Participation was a 'fail' at the principle level; likewise interest representation as a criterion. Resources was also the lowest scoring indicator (1.94), and accountability the highest (2.86). Problem-solving, inclusiveness and agreement also received 'fail' ratings; in the case of inclusiveness, contributing to the 'fail' for interest representation – along with equality – at the criterion level. These results

Table 8.4 EU perspectives on the quality of climate governance by level of policymaking activity

Principle	1. Meaningful Participation Maximum score: 25 Minimum: 5							
Criterion	*1. Interest representation* *Maximum score: 15* *Minimum: 3*				*2. Organizational responsibility* *Maximum score: 10* *Minimum: 2*			Principle Score
Indicator	**Inclusiveness**	**Equality**	**Resources**	*Criterion Score*	**Accountability**	**Transparency**	*Criterion Score*	
Intergovernmental	**2.33**	**2.06**	**1.94**	***6.33***	2.86	2.64	*5.5*	**11.83**
Regional	2.67	**2.42**	**2.06**	***7.15***	3	2.69	*5.69*	12.84
National	3.04	2.68	**2.32**	*8.04*	3.16	2.95	*6.11*	14.15
Other	**2.3**	**1.7**	1.33	*5.33*	**2**	**1.83**	*3.83*	**9.16**

Principle	2. Productive deliberation Maximum score: 30 Minimum: 6								
Criterion	*3. Decision-making* Maximum score: 15 Minimum: 3				*4. Implementation* Maximum score: 15 Minimum: 3				Principle Score
Indicator	**Democracy**	**Agreement**	**Dispute settlement**	*Criterion Score*	**Behavioural change**	**Problem-solving**	**Durability**	*Criterion Score*	
Intergovernmental	2.85	**2.38**	2.62	*7.85*	2.62	**2.31**	2.62	*7.55*	15.4
Regional	2.93	2.69	2.75	*8.37*	2.85	**2.36**	2.62	*7.83*	16.2
National	3	2.73	2.86	*8.59*	3.13	2.53	3	*8.66*	17.25
Other	**2.2**	2.5	**2**	*6.7*	**1.75**	**1.75**	**2**	*5.5*	**12.2**
Total (out of 55)									
Intergovernmental									**27.23**
Regional									29.04
National									31.4
Other									**21.36**

Note: Light grey represents the highest score; dark grey the lowest; numbers in bold are below the threshold value of 50 per cent; average results from each sector.

could be read as a reflection on the dominance of the nation-states in the high-level policymaking that occurs under UNFCCC. On this view, negative ratings could be interpreted as a perception of exclusiveness and inequality *contra* member states.

'Other' received the lowest score (39 per cent); both principles were a 'fail', as were all criteria. Resources was also the worst performing indicator, with the lowest 'fail' rating overall (1.33). With the exception of agreement (the highest rated indicator), which was a borderline 'pass', all remaining indicators were a 'fail'. While these results reflect views on a disparate range of policymaking localities, some credence should be given to these perspectives, given the relatively high numbers of respondents (31). In this case, the results would appear indicate a general preference for more state-oriented venues.

Table 8.5 Commentary on Results

In contrast to those from the EU, the non-EU results demonstrate a preference for the intergovernmental level of climate policymaking but the score was a 'fail' nevertheless (49 per cent). Participation did not meet the threshold, while implementation was the only 'pass' at the criterion level. Resources was the lowest rated 'fail' indicator (1.58) and inclusiveness the highest (2.79). Other indicators that did not meet the threshold were transparency and accountability, as well as dispute settlement and agreement (explaining why their related criteria were both a 'fail'). No respondents rated any indicator at 3 or above at this or any other level. This is another poor performance, especially given the results for transparency, accountability, agreement and dispute settlement (recurring themes across this set of results). The fact that this level received the highest score amongst non-EU respondents may reveal a greater of degree of confidence in intergovernmentalism than other avenues for policymaking, but the results remain unimpressive.

National level policymaking was a close second (also 49 per cent, with rounding), replicating the principle and criterion results of the intergovernmental level, with implementation likewise receiving a (slightly lower) 'pass'. Democracy was the highest rated indicator overall (2.85) and resources the lowest overall (1.54); the other indicators to also receive a 'fail' were accountability (the second lowest indicator for this level) and transparency (explaining the criterion score). Problem-solving was also a 'fail', although behavioural change and durability were a 'pass', thereby resulting in a low 'pass' for the criterion overall. Agreement and dispute settlement did not meet the threshold, also following the intergovernmental results but with a slightly higher score for democracy (the second

Table 8.5 Non-EU perspectives on the quality of climate governance by level of policymaking activity

Principle	1. Meaningful Participation Maximum score: 25 Minimum: 5							
Criterion	*1. Interest representation* *Maximum score: 15* *Minimum: 3*				*2. Organizational responsibility* *Maximum score: 10* *Minimum: 2*			Principle Score
Indicator	**Inclusiveness**	**Equality**	**Resources**	*Criterion Score*	**Accountability**	**Transparency**	*Criterion Score*	
Intergovernmental	2.79	2.54	**1.58**	***6.91***	**2.33**	**2.46**	***4.79***	**11.7**
Regional	**2.34**	**2.36**	**1.59**	***6.29***	**2.33**	**2.48**	***4.81***	**11.1**
National	2.85	2.51	**1.54**	***6.9***	**2.32**	**2.39**	***4.71***	**11.61**
Other	**2.26**	**2.42**	**1.7**	***6.38***	**2.18**	**2.24**	***4.42***	**10.8**

Table 8.5 (continued)

Principle	2. Productive deliberation Maximum score: 30 Minimum: 6								
Criterion	*3. Decision-making* Maximum score: 15 Minimum: 3				*4. Implementation* Maximum score: 15 Minimum: 3				Principle Score
Indicator	**Democracy**	**Agreement**	**Dispute settlement**	*Criterion Score*	**Behavioural change**	**Problem-solving**	**Durability**	*Criterion Score*	
Intergovernmental	2.55	**2.43**	**2.36**	***7.34***	2.74	2.57	2.78	*8.09*	15.43
Regional	2.64	**2.33**	**2.27**	***7.24***	2.67	**2.21**	**2.42**	***7.3***	**14.54**
National	2.76	**2.34**	**2.35**	***7.45***	2.7	**2.39**	2.7	*7.79*	15.24
Other	2.69	**2.38**	**2.07**	***7.14***	**1.93**	**1.93**	**2.07**	***5.93***	**13.07**
Total (out of 55)									
Intergovernmental									**27.13**
Regional									**25.64**
National									**26.85**
Other									**23.87**

Note: Light grey represents the highest score; dark grey the lowest; numbers in bold are below the threshold value of 50 per cent; average results from each sector.

highest indicator at this level). Even though decision-making was scored the highest of all the levels, this was not enough to avoid a (borderline) 'fail' – not a very positive sign. However, both democracy and inclusiveness were rated higher than at the intergovernmental level. These results would further appear to demonstrate a lower level of satisfaction regarding the efficacy of national level activities when compared to the EU results, with a lacklustre performance for many aspects of governance.

The regional level was the third highest performer (47 per cent). Neither of the principles nor the criteria met the threshold. Resources was again the lowest scoring indicator (1.59) and behavioural change the highest (2.67); democracy was the only other indicator to 'pass'. This is not a good performance and the results appear to show that climate policymaking at the regional level was not perceived especially favourably.

'Other' levels of policymaking generated the lowest scoring results with a 'fail' overall (43 per cent), including both principles and all criteria. Resources was the lowest rated indicator (1.7), while democracy was highest (2.69) and the only indicator to exceed the threshold.

Table 8.6 Commentary on Results

The results from employers' organizations showed the governance of climate policymaking to be highest at the national level (53 per cent). Both principles were the highest scoring overall but participation did not meet the threshold value due to the borderline 'pass' for organizational responsibility and the 'fail' for interest representation. Implementation was the highest scoring criterion overall and, along with decision-making, exceeded the 'pass' threshold. Resources was the lowest rated indicator overall (1.2) and a 'fail' along with transparency. All other indicators were a 'pass', with some positive ratings for inclusiveness, the highest indicator overall (3.38), and also for behavioural change and democracy. This is a relatively sound performance but, on the basis of the rating for transparency and a comparatively weak performance for accountability, it may be possible to infer a general concern regarding the extent to which it respondents deemed it possible to determine what was really going on in domestic-level climate negotiations. This impacts negatively on the degree to which the participation of employers' associations might be determined as being meaningful. Interestingly, the comparatively high rating for inclusiveness offsets the very poor performance of resources, and may demonstrate a perception that employers' organizations felt included in policymaking, even if equality was not rated so highly. Nevertheless, the good results for decision-making and implementation could be interpreted as revealing a degree of confidence in national level processes.

Table 8.6 Employer association perspectives on the quality of climate governance by level of policymaking activity

Principle	1. Meaningful Participation Maximum score: 25 Minimum: 5							
Criterion	*1. Interest representation Maximum score: 15 Minimum: 3*				*2. Organizational responsibility Maximum score: 10 Minimum: 2*			Principle Score
Indicator	**Inclusiveness**	**Equality**	**Resources**	*Criterion Score*	**Accountability**	**Transparency**	*Criterion Score*	
Intergovernmental	3.27	2.78	**1.25**	*7.3*	**2.44**	**2.44**	***4.88***	**12.18**
Regional	2.82	2.75	**1.29**	***6.86***	2.63	**2.25**	***4.88***	**11.74**
National	3.38	2.64	**1.2**	*7.22*	2.55	**2.45**	*5*	**12.22**
Other	**2.43**	2.5	**1.5**	*6.43*	**1.8**	**1.6**	*3.4*	**9.83**

Principle	2. Productive deliberation Maximum score: 30 Minimum: 6								
Criterion	*3. Decision-making* Maximum score: 15 Minimum: 3				*4. Implementation* Maximum score: 15 Minimum: 3				Principle Score
Indicator	**Democracy**	**Agreement**	**Dispute settlement**	*Criterion Score*	**Behavioural change**	**Problem-solving**	**Durability**	*Criterion Score*	
Intergovernmental	2.83	2.71	2.71	*8.25*	3	2.86	2.71	*8.57*	16.82
Regional	3.17	2.67	2.67	*8.51*	3	2.71	**2.33**	*8.04*	16.55
National	3	2.63	2.63	*8.26*	3.25	2.63	2.75	*8.63*	16.89
Other	3	**2.33**	**2**	*7.33*	**1.67**	**1.67**	**1.33**	***4.67***	**12**
Total (out of 55)									
Intergovernmental									29
Regional									28.29
National									29.11
Other									**21.83**

Note: Light grey represents the highest score; dark grey the lowest; numbers in bold are below the threshold value of 50 per cent; average results from each sector.

The intergovernmental level was a close second (also 53 per cent). As with the national level, participation did not meet threshold, this time on account of a borderline 'fail' for both organizational responsibility and interest representation (which achieved a slightly higher score at this level). Resources was the lowest rated indicator (1.25) and inclusiveness also once again the highest (3.27). Accountability and transparency performed poorly (two 'fails' with the same rating), while all other indicators met the threshold, with the results mirroring those at the national level, albeit slightly lower. Here, the trend regarding accountability and transparency was articulated more strongly with lower ratings. It is possible that perceptions regarding the national (governmental) level have been carried over to the intergovernmental level, where member states on occasion negotiate climate policy behind closed doors and cannot be held to account by their citizenry in this extra-national context. It is worth noting that equality fared slightly better at this level, which, along with inclusiveness, again partially compensates for the low rating for resources.

The regional level of policymaking achieved the third highest score (51 per cent) but the difference remained close. The results at both the principle and criterion levels were similar to the intergovernmental level, with participation, interest representation and organizational responsibility receiving a 'fail'. This time decision-making was the highest scoring criterion, and democracy was the highest scoring indicator (3.17). Once again, resources was the lowest 'fail' (1.29). At this level, both transparency and durability failed to meet the threshold. The other indicators were a modest 'pass', with only one other indicator in the '3' band (behavioural change, the second highest rated indicator). These results would appear to confirm the trend regarding issues of organizational responsibility in climate policymaking, although it should be noted that accountability was a 'pass' at this level.

'Other' levels of policymaking received a 'fail' (40 per cent), although the variety of venues (public, private, local, and so on) makes it difficult to draw any firm conclusions. Both principles did not meet the threshold, with the lowest scores overall, nor did any of the criteria (also with the lowest scores overall). For the first and only time across all the results, resources was not the lowest rated indicator; this fell to durability (1.33). Democracy was the only indicator to merit a 'pass', with a relatively good result (3).

Table 8.7 Commentary on Results

The trade union results also delivered the highest score to climate policy at the national level (51 per cent). Both principles met the threshold level

Table 8.7 Trade union perspectives on the quality of climate governance by level of policymaking activity

Principle	1. Meaningful Participation Maximum score: 25 Minimum: 5							
Criterion	*1. Interest representation* *Maximum score: 15* *Minimum: 3*				*2. Organizational responsibility* *Maximum score: 10* *Minimum: 2*			Principle Score
Indicator	**Inclusiveness**	**Equality**	**Resources**	*Criterion Score*	**Accountability**	**Transparency**	*Criterion Score*	
Intergovernmental	**2.47**	**2.31**	**1.9**	***6.68***	2.63	2.67	*5.3*	**11.98**
Regional	2.5	**2.42**	**1.94**	***6.86***	2.58	2.68	*5.26*	**12.12**
National	2.83	2.57	**1.92**	*7.32*	2.65	2.65	*5.3*	12.62
Other	**2.29**	**2.14**	**1.67**	***6.1***	**2.29**	**2.35**	***4.64***	**10.74**

Table 8.7 (continued)

Principle	2. Productive deliberation Maximum score: 30 Minimum: 6								
Criterion	*3. Decision-making* Maximum score: 15 Minimum: 3				*4. Implementation* Maximum score: 15 Minimum: 3				Principle Score
Indicator	**Democracy**	**Agreement**	**Dispute settlement**	*Criterion Score*	**Behavioural change**	**Problem-solving**	**Durability**	*Criterion Score*	
Intergovernmental	2.73	**2.44**	**2.44**	*7.61*	2.7	**2.44**	2.81	*7.95*	15.56
Regional	2.8	2.52	**2.44**	*7.76*	2.76	**2.21**	2.59	*7.56*	15.32
National	2.83	**2.43**	**2.44**	*7.7*	2.72	**2.33**	2.78	*7.83*	15.53
Other	2.59	2.5	**2.13**	*7.22*	**2**	**2**	**2.29**	***6.29***	**13.51**
Total (out of 55)									
Intergovernmental									27.54
Regional									**27.44**
National									28.15
Other									**24.25**

Note: Light grey represents the highest score; dark grey the lowest; numbers in bold are below the threshold value of 50 per cent; average results from each sector.

and, in the case of participation, achieved the highest score. However, it should be noted that this result for participation was borderline and only just exceeded the 50 per cent threshold. Three criteria were a 'pass', with decision-making and organizational responsibility also receiving the highest scores at the criterion level (organizational responsibility was scored equally highly at the intergovernmental level). However, it should be noted that while a 'pass', these results are not a resounding endorsement of the transparency and accountability of policymaking, nor related decision-making. Interest was representation received 'fail', albeit the highest scoring 'fail'. Resources was again the lowest indicator (1.92) and inclusiveness and democracy the highest (2.83). Democracy enabled the criterion of decision-making to exceed the threshold, but it should be noted that both agreement and dispute settlement were a 'fail', as was problem-solving (the lowest rating at this level of policymaking). These results are moderately satisfactory but not impressive, and are slightly lower than those from employers, although it is interesting to see a higher score for transparency. In further contrast, the low ratings for agreement and dispute settlement are to be considered.

Like employers, the intergovernmental level of policymaking was the next best performer; however, it was still only a borderline 'pass' (50 per cent). At the principle level, deliberation met the threshold, with the highest result, while participation did not. Three criteria met the threshold but interest representation was a 'fail' – a stark contrast to employers. Resources was the lowest rated indicator (1.9) and durability the highest (2.81). Four other indicators were a 'pass': democracy (the second highest rating), behavioural change, accountability and transparency. In the case of decision-making, it was only the comparatively high score of democracy that prevented this criterion from failing to meet the threshold, as both agreement and dispute settlement were a 'fail'. Problem-solving was also a 'fail'. These results are not a ringing endorsement of intergovernmental policymaking.

Regional level policymaking was third, but with slightly worse results than the intergovernmental level (50 per cent). Similarly, deliberation met the threshold at the principle level while participation did not. Three criteria also met the threshold, with interest representation the 'fail'. Resources was the lowest rated indicator and a 'fail' (1.94), with democracy the highest 'pass' (2.8). Other indicators to meet the threshold were behavioural change (the second highest indicator), followed by transparency, durability, accountability, agreement and inclusiveness (in this case, on the borderline). However, the 'fail' for equality, and the borderline 'pass' for inclusiveness might appear to indicate a negative perception about interest representation.

'Other' was the least performing level of policymaking activity (44 per cent) but the previous caveats about the disparate nature of this category need to be restated. Neither deliberation nor participation met the threshold, nor did any criteria: these were all the lowest results overall. Resources was also the lowest rated indicator overall and a 'fail' (1.67), Democracy was the highest 'pass' (2.59), with agreement the only other (borderline) 'pass'; all other indicators did not meet the threshold. After resources, behavioural change and problem-solving were the second lowest scoring indicators overall (2), perhaps a reflection of the fact that respondents' thought that governmental processes (ought to) have a greater ability to implement climate policy.

OBSERVATIONS ON THE GOVERNANCE ANALYSIS FROM SURVEY RESPONDENTS

Commentary from Employers

Given the number of respondents, only a few comments were offered. In relation to inclusiveness, one respondent questioned whether governments were prepared 'to commit to climate change policies', while another was of the view that those interests who wanted to introduce new initiatives were 'often overlooked' because they were seen as being 'outside mainstream thinking'. For this respondent, there was a knock-on effect in terms of equality, because it meant that if a particular initiative was not 'mainstream' it was 'not rated as important'. Two further respondents commented briefly on resources. One noted that in terms of support from climate policymaking forums they received 'actually, none'. Another respondent commented that they did 'not expect any contribution'. Two further respondents provided comments on transparency, both of which were negative. One believed that transparency lost out to those issues that were 'politicized', consequently 'fact gets lost in the process'. Another respondent commented that the policymaking they had been involved in was 'not transparent'. They felt that their presence 'was just to fulfill [a] requirement'. In terms of dispute settlement, this same respondent considered that is was the 'people in authority' who had the 'greater say'. This led them to conclude that as far as problem-solving was concerned, 'policy decision-makers are not interested'.

Commentary from Unions

With larger numbers of respondents, there were correspondingly more comments from this sector. For inclusiveness, comments fell into three main categories: the extent to which unions were or were not included in formal policymaking processes; the difficulties experienced in trying to get climate change issues recognized inside and outside the labour movement; and what the broader contribution of unions to climate-related policymaking should be. In terms of being included in governmental processes, one respondent commented that they had 'no formal role at any level', despite the fact that their union had 'participated in community level climate change engagement at the local, State and (to a limited extent) national level'. Another respondent complained that:

> Organized labour and NGOs have struggled to get Just Transition or even human rights impacts in the language of UNFCCC documents. We have a toehold in the COP21 Paris Agreement. At the national level we have been a significant contributor to climate policy of the [previous government], but still have to struggle there. With the [current government], it is mostly a closed door for unions.

Concerning the extent to which unions were included in climate change issues, one respondent noted that there was generally only a 'low recognition of the relevance of trade unions outside of those covering directly impacted workers (for example, coal and energy)'. Their union included 'climate scientists' in particular, but despite this they had found it necessary 'to continually assert our relevance and informed contributions'. They had found that 'the issue for us is with government, business and NGOs as well as other unions'. Another respondent from a national union described how climate issues were handled internally, explaining that local branches 'undertake their own activities to varying degrees, our larger branches engage more directly with members by holding annual conferences on climate change'. One union member commented that they did 'not feel involved in any climate policy decisions. The only participation is my own commitment to change the policy in my home and work offices'. With regard to the role of CSOSs, one respondent thought that: 'more formal engagement with unions and civil society is needed at national level'. Another respondent echoed this sentiment, pointing to the 'need for more public awareness that would ensure that decisions are implemented and monitored'.

In commenting on issues of equality, there were a few observations about the bias of policy negotiations. Although, as one respondent put it, 'all participants are treated with respect' they were aware that 'commercial and political interests skew attitudes'. Another was more frank, suggesting that processes were 'dominated by single-issue pressure groups with a contempt for working people'. A third respondent, reflecting a similar view recognized that 'as a trade union, we are not a negotiating party. Given that, the parties (to the treaty) tend to listen most to those shouting the loudest – often business lobbies'. Another respondent from a climate vulnerable country saw the problems that can arise from unequal treatment of unions in society:

> The importance of public sector unions' participation is recognized by certain bodies. Nonetheless, such recognition is wanting in other strategic bodies that tackle matters concerning climate change, disaster risk reduction and management [DRRM] and the like. For instance, in the drafting of the [national] DRRM Plan, public sector unions were not represented. This matter is highly important since distinct inputs can be contributed by the public sector union not only because of the involvement and responsibility of its members in matters of public services but also because most often than not, these same members are themselves victims of disasters and/or the consequences of climate change. Union participation at all levels should be advanced.

Lastly, one union representative expressed concern about the unequal treatment given to different policy-related impacts of climate change, notably, given the topic of this book, on workforce issues:

> Everyone says something about jobs, but no one integrates labour structural adjustment into their climate policies and programs. We have rhetoric without substance from most players – from industry through all major parties to green groups.

Issues around resourcing attracted a number of responses, which can be broadly divided into three topics: observations about the lack of resources to participate in climate-related policy; how different unions responded to this lack; and, by way of contrast, commentary on the resources that other sectors had at their disposal. Several respondents commented that as trade unions, they did not receive any external funding for their participation, as one noted 'in COPs or for our other activities on climate'. Another explained that their union did not receive 'any external resources for climate change activity'. One respondent provided some historical context:

> I don't think unions in [country] have ever received [money] for any work on climate policy or programs or to participate in any event. Way back in the 1989–92 period the [government agency] did provide travel costs to unions for participation, but nothing to fund policy development, while green groups got hundreds of thousands of dollars. The most assistance that has ever been given at international level is the rare inclusion of unions (alongside industry

> and green groups) in government delegations to UNFCCC events where NGO access was highly restricted.

Sources of income for participation varied. One union member explained that they were 'expected to self-fund' while another complained that they were 'always expected to pay up – and support member participation and also help out NGOs'. Some respondents explained that their sources of income for participation were derived 'internally' and, as another explained, 'we finance all our activities on our own funds (workers' fees)'. The organizer of an environmental action group within their own union explained that this was 'a member driven group and as such attracts no allocated funding or administrative support. It is facilitated in my role as Vice President, and I manage the requirements within my own resource boundaries'. Another respondent commented on the resources for government agencies in comparison to civil society, and made some suggestions:

> Government representatives participating in discussions/forum/workshops and like sessions are fully funded by government. On the other hand, trade unions have to shoulder costs. Hence, if resources are limited and/or the activity [is] not covered by the approved program of action of the union it is most likely that participation is also constrained. Recognising the importance of trade union inputs, mechanisms should be in place to allow a fully sponsored participation. After all, each government agency has its distinct programs, fully funded, from which such full sponsorship can be sourced.

In providing comments on matters of accountability, it was interesting to see that two respondents contrasted the UNFCCC negotiations favourably in comparison to international trade negotiations. One noted that UNFCCC did 'at least provide for civil society presence and input', while the second respondent was of the view that in comparing the two, 'climate negotiations are transparent, pluralistic, open and democratic'. But this same respondent was 'not convinced that trade negotiations might really constitute a standard, though'. At the national level, the trade union representative from a climate-vulnerable country and who had previously commented on their country's DRRM plan, linked it to accountability:

> The Plan is there, but is there a tool to monitor if indicators are achieved and/or who are the persons/offices responsible? Sadly, there is none. Hence, the urgency for the drafting of a monitoring tool that will also indicate responsible offices/persons as a clear way of pinpointing accountability. The trade union however, as earlier mentioned, was not given the opportunity to sit in the drafting of the plan. In the [confederated trade union] bodies where [this respondent's union] sits, inputs are utilized as part of the policy/action recommendations, which these bodies consistently push and campaign for.

Given the strong conceptual linkages to accountability, several of those respondents who commented on transparency referred back to the previous question. Two further comments were offered. One union member thought that the transparency of climate policymaking at the national level was 'reasonable' but indicated that because the level of participation afforded to them by their national government was 'low', their ability to comment was 'limited'. One final respondent, discussing the nature of policymaking, felt that 'there is always an agenda by those that run the conferences'.

Three respondents commented on issues relating to democracy. On a national level, the first was of the view that their government was not 'very interested in climate change policy and has until recently been winding climate policy back'. In the light of recent developments, they thought this was 'now being partially rectified, however progress is insanely slow. Climate change policy seems quite separate to any democratic processes in place at the national level'. Two respondents active at multiple levels, including at the intergovernmental level, questioned the democratic legitimacy of states active within the climate regime. One observed that:

> UNFCCC and other intergovernmental processes are democracy of a kind. The UNFCCC requires consensus of all parties – which gives equal weighting to the population of China and Tuvalu! . . . However, it is pretty clear that governments listen to banks, mining companies and other big business more than civil society.

The third also wondered how much negotiating parties took the views of their citizenry into account, reflecting that: 'democracy refers to the people. Parties to the treaty are states. Do states' or nations' positions really reflect people's interests in a faithful manner?'

Comments on the effectiveness of agreements within climate policy ranged from observations about internal union processes and community level activities as well as national and international processes. Internally, one respondent active on environmental issues contended that within their union the 'process of policy formulation, consideration and adoption was transparent, inclusive and democratic'. A further respondent linked their own union's internal processes to the making of broader trade union policy, noting that: 'while resolution/s can be used to express the overall perspective and collective position of participating organizations/bodies, there is still a need to cascade such position up [sic] to the grassroots for wider support'. Another thought that the community action had proven to be one of the most effective avenues for climate action and that 'community voices, coordinated through groups such as 350.org, appear to have impacted the negotiations, however, our country as a whole has provided little if any avenues for contribution to the public debate'. In making the link between union policy and government, one respondent questioned which was more effective, given that:

> The government has the ear of climate change deniers. We wonder if there is a genuine desire to address climate change policy or is it a token gesture . . . our own effectiveness is related to [country] branches communicating with members about climate change, and finding out what is going on [in the union branches] and initiatives our members make in their workplaces.

Looking at the intergovernmental level, one of two respondents thought that the 'UNFCCC talks do not advance fast enough to counter climate change', while the second considered that the 'UNFCCC approach is the broadest, but it is clear that major progress will depend on agreement between the largest countries and largest emitters (roughly the same group)'. This respondent pointed to the importance of other forums, 'like US–China deals, the Major Economies Forum and the G20'. At the national level, they thought climate policy had become 'highly polarized' and made a plea for 'climate policy to become bipartisan to provide certainty over a period of decades'.

In relation to dispute settlement, respondents were either unaware of specific disputes, or had negative perspectives. One was of the view that 'avoidance of areas of dispute seems the usual way out – and then finding lowest common ground'. Another, who commented previously about their national government's hostility to action on climate change, believed that 'the whole question is still in dispute, especially within the current government, and even within the opposition in terms of being politically committed to take meaningful action'. One of the intergovernmental level commentators made the pertinent observation that 'the UNFCCC process does not really have dispute settlement or enforcement mechanisms. In fact the process of target-setting involves each country nominating its own targets unilaterally'. In their own country, this respondent thought climate policy 'has been a battlefield with little dispute resolution'.

Views on the likelihood of climate policy changing the behaviours that have contributed to historical anthropogenic emissions were both varied and nuanced. One respondent stated that they were 'always confident that exposure to the debates and informed discussion changes behaviours', a view shared by another who believed that 'awareness raising' had 'provided a platform to base future action upon'. A third respondent agreed, with the proviso that 'if able to be involved, I believe I can change behaviour'. A further respondent was less optimistic, thinking that the chances of changing behaviour were 'very doubtful', but they did think it might be possible if their national confederation of trade unions were to 'take climate change on, on behalf of their affiliates (our union being one) in a big way'. Another respondent expanded on the mechanics of gaining such traction at the confederated trade union level. They had a body in their country that brought a range of different unions together to develop common positions but in order to 'achieve its fruits' it needed 'tangible follow-through programs for policies'. They thought that the 'incorporation of policy recommendations . . . should be harmonized and complemented' with other climate-related programmes. They had confidence behavioural change could be 'gotten and reinforced with the right mechanisms, not solely through policy negotiations. And it is something that the government, the private sector and trade unions can jointly work for'. Finally, one of the union representatives active internationally and nationally considered that globally there is 'momentum building for action, but the question is whether the pace will ever be enough . . . At the national level, government policy has tended to lag behaviour change, though the picture is murky'. They pointed to the fiscal instruments and incentives developed by government to encourage solar uptake, but they concluded that at this point, it was now 'market economics and climate awareness that is doing more of the work'.

Respondents' views were similarly varied on the extent which climate policies would actually solve the problem of climate change. One thought there was 'lots of politics, lots of lobbying' leading to actions that, according to a second respondent, were 'largely symbolic'. A third respondent commented along similar lines that: 'while there is satisfaction in talking and settling upon conservative outcomes, little will change'. A fourth respondent believed that with their country 'being a signatory to the Paris Agreement is as close as it's come'. But they also thought 'the action, which should follow, is not evident' and that 'the political cycle is one of the barriers to ongoing commitment and follow through action'. One final respondent summed up the possibilities:

> There will be progress towards 'solving' climate change, but it is an open question as to whether the pace will be enough. It seems pretty clear that there will be an overshoot in terms of emissions not reducing fast enough to proven warming above 2 degrees, but the solutions to reduce emissions more rapidly may occur and be deployed.

In their comments on durability, one respondent thought this was 'the key issue'. They explained that their 'interventions have been in the national political processes focused upon legislation and government policy' but 'durability is just not there, as there is not a consensus or way to get there yet'. Noting the need for flexibility and adaptability, a further respondent made the logical point from an organized labour perspective that 'as for all policies, there will be a review, at which time the stance may be altered depending on current membership views'. This was because, according to a second respondent, 'the perspective and priority of the current leadership also matters'. Another respondent made the insightful observation that 'the whole debate has been going on for over 20 years, so it's had longevity, but not resilience, flexibility, or adaptability'. This perspective was reflected by one of the union representatives active at the international and national levels. They concluded that 'durability is improving at the UNFCCC level' but that nationally there needed to be 'a complete revamp of climate policy and politics'.

Commentary from Other Respondents

Other respondents also made some useful contributions about the governance of climate policymaking. Concerning inclusiveness, one respondent who is active in the area of 'civic education' felt that 'negotiations seem limited to a handful of 'special' players'. This view was shared by another respondent who had 'attended one policy think tank organized by a government department' but considered that it had been 'farcical'. A third respondent commented on inclusiveness and suggested that policymaking 'should include those most impacted, be diverse and include the voices of those who bear the brunt of climate change', also adding that such forums should 'be linked to anti austerity, anti-racist, anti-fascist and migrant rights' discussions and campaigns'. The final respondent to comment thought 'access to consultations' through avenues such as municipal/mayoral planning had afforded them 'no real role in either policymaking, or formally holding bodies to account'. No one commented on issues of equality, but regarding resources this same respondent felt that their government was 'specifically excluding' stakeholders by 'strictly limiting resources – for example travel and subsistence to attend meetings' meaning it was necessary for them to 'fund their own way'.

This same respondent commented on both accountability and transparency at the national level, noting in terms of the former that: '"what is said in the room stays in the room" is a key principle, and if you exclude those who may challenge you from the room, then there is limited accountability'. Concerning transparency, they went on to add that they were 'especially unable to get at information as to what happens when the government fails to meet legally set EU pollution standards'. This led them to conclude that this was a strategy to allow government to 'negotiate around these [standards] in confidence'. They also questioned the democratic nature of climate-related decision-making in their country, making the observation that 'even in its broadest sense there is little accountability'. They did not consider any of the agreements made to be especially effective because 'outcomes are obscure and rarely policed. Fines are rarely imposed. Targets [are] never met and budgets for greater sustainability rarely, if ever, adapted to facilitate the changes'. As far as dispute settlement was concerned, they made the point that they 'never' got to participate in disputes because their country did not 'follow the social partnership model, and the institutions that do nominate up to the EU' did not 'engage downwards' to stakeholders on 'environmental issues'.

Two respondents provided observations on matters around implementation. In terms of the impacts of climate policy on changing behaviour one respondent was pessimistic, commenting that: 'the exclusive nature of the negotiations and the reasons for this – putting vested political and capital interest first – excludes confidence in genuine change'. The second noted that this was 'entirely dependent on the extent of adoption of the policy that has been put in place'. They pointed to recent developments by some companies to disclose their climate-related activities at annual general meetings and added that the Carbon Disclosure Project (CDP) 'has stated that if widely adopted, it would be a game changer in driving corporate change in relation to climate change'. Both commented on the capacity of policymaking to solve the problem of climate change. The first made the point that 'intergovernmental and regional policies cannot work without regulation and enforcement, and the same bodies are unwilling and or unable to police themselves'. In relation to both problem-solving and durability, the second respondent reiterated their previous comments about the need to first formulate policy and then adopt it.

ANALYSIS AND CONCLUSIONS

The results from employers' organizations were modest. However, it might be concluded that they demonstrate a general satisfaction concerning interest representation at the national, intergovernmental and regional levels but dissatisfaction with transparency and, to a certain extent, accountability. Clearly, there were poor ratings for resources. This is perhaps not surprising in view of the fact that many decision-making venues do not resource stakeholders to attend (or restrict resourcing to developing country participants, for example). By nature, organizations such as employers' associations represent others and are often afforded the capacity to do so by their members (through, for example, fees and subscriptions). But this may not be the case for some (such as those in less developed countries). In the case of other comparative studies of UNFCCC-related initiatives (REDD+, CDM) undertaken by the researchers, resources have also been consistently the lowest rated indicator (Cadman et al., 2016; Maraseni and Cadman, 2015). It would be expected that employers' organizations might have the economic capacity to attend meetings themselves but there are other kinds of resources, such as the provision of technical or institutional support, that may be absent. If so, the responsibility is on the institutions developing climate policy to provide these. What is clear in the results is that climate policymaking was not rated as particularly transparent at any level, and there were reservations about accountability at the intergovernmental level as well. As UNFCCC is arguably the most important climate policymaking venue of all, this is of concern. Given the number of respondents, care should be exercised in viewing these results as a definitive expression of employers' views, however the trends are certainly interesting.

Trade unions were really only satisfied with policymaking at the national level but even here there were problems. Problem-solving and dispute settlement were consistent 'fails', while equality and inclusiveness were not especially high performers; nor was agreement. Particular attention should be given to interest representation, which did not meet the threshold at any level and stands in contrast to the employer results. While the numbers of unions responding are not large given the total global labour movement, the results could nevertheless be seen as being at least partially indicative of this sector's perspectives on the governance quality of climate-change related policy. Trade union respondents appear to feel excluded from policymaking at the intergovernmental level and have issues regarding their equality of treatment.

EU respondents viewed their role in policymaking at both the national and regional level positively, but issues around equality and problem-solving were recurring themes at the intergovernmental and regional levels. The relatively high ratings for both accountability and transparency are also favourable, but it should be noted that resources was also identified as a problem at all levels of policymaking. The low results for the governance quality of policymaking at the intergovernmental level, which may be connected to a perceived lack of interest representation, should be of concern. Nevertheless, on the basis of the results, the conclusion might be drawn that these respondents viewed their membership of the EU as making a positive contribution to their involvement in the governance of climate-related policymaking.

This was not the case for respondents outside the EU, and those results show a far less positive attitude regarding involvement in climate policymaking. There were some mitigating factors. Inclusiveness and equality were rated favourably at the intergovernmental and national levels, as was durability, and behavioural change was a 'pass' at the intergovernmental, national and regional levels. These latter two results may demonstrate that the implementation of climate policy was perceived as having some impact on reducing emissions, even if this was not that successful in solving the broader problem of climate change. However, there were also significant weaknesses at all levels. Accountability and transparency were seen as lacking at all levels, which is deeply concerning. Democracy fared comparatively better being the only indicator to 'pass' at all levels of policymaking, but this was offset by the fact that both agreement and dispute settlement were also universal underperformers. The other negative message was that respondents were clearly of the view that policymaking institutions did not provide sufficient resources to assist in their participation in policymaking.

A number of conclusions can be drawn from the survey. There appears to be more confidence in climate policymaking at the national level than via UNFCCC. This is partially related to concerns around access and influence, and accountability and transparency, which are more readily addressed at the ballot box than at the global climate talks if stakeholders are dissatisfied. Greater emphasis must be placed on transparency of the climate talks themselves, not just about transparency in relation to countries reporting on their emissions reduction activities or which country has given how much money to which other country or initiative (Decisions 85–99 of the Paris Agreement). COP 21 in Paris was heavily criticized by CSOs, who argued 'negotiating behind closed doors undermines the ability of civil society to ensure the accountability of governments and the UNFCCC process' (ECO, 2015).

But equally important, if not more so, is this extent to which the availability of resources for CSOs at the intergovernmental level impacts their ability to represent their interests effectively. Participation in international policymaking is expensive, given the costs of travel and accommodation. Effort has been put in at the climate talks to provide wireless Internet and there are some opportunities for getting hold of hard copies of negotiating text. However, it is undeniable that there is much better support for member states that are parties to the Convention than CSOs. Given the significance that the Paris Agreement places on the role of 'non-Party stakeholders', as they are referred to (Decisions 134–137), and the commitment to capacity building within the text (Decisions 72–84), it is vital that better provisions are made (in the texts of future agreements, if necessary) to allocate more resources to CSOs to attend negotiations, not just Parties. It is speculation, but possible, that unions may feel more excluded at the intergovernmental level than business interests because their capacity to attend is more constrained. It should be emphasized here that the survey question referred to the extent to which policymaking forums provided resources to survey respondents. Here, there is clearly a deficit and, in the case of unions at the very least, and most likely for other sectors as well, there is a need for greater allocation of resources by policymaking forums for CSOs to attend. Given the comparatively low numbers of trade union representatives in climate negotiations historically (consistently less than 1 per cent of registered attendees over time), it is possible that TUNGOs are losing out compared to ENGOs or BINGOs (15 per cent and 9 per cent respectively); this might explain the contention (Cabré, 2011) that sectors such as business and environmental interests 'continue to dominate civil society participation at UNFCCC events' (p. 20). In this case, providing resources for under-represented sectors is even more important.

In addition to this, it must be reiterated that CSOs are often deliberately excluded from intergovernmental negotiations, with a consequential impact on both transparency (since it is not possible for them to determine if their views are being represented) and accountability (since Parties cannot be held to account for any misrepresentation that might occur). Interestingly, it appears that it is not the case that non-state actor participation 'clutters up' the negotiating space, thereby undermining efficiency. Rather it is the political dynamics, and historical institutional factors that lead to negotiations behind closed doors, and ironically, not the sensitivity of particular agenda items (Nasiritousi and Linnér, 2016). If this is the case, governmental Parties need to yield political space to other actors and the UNFCCC needs to increase institutional efforts to develop better modes of multi-stakeholder participation.

9. Conclusions

This book is motivated by the concern that climate change policy generally does not reflect the impacts of those policies on the labour market and that there is a need to inform labour market practitioners of the impact that climate change policy will have on their spheres of interest. The book also sought to test how an appropriate theoretical framework can guide the process of the development of climate change policy.

In responding to its central research questions regarding the impacts of climate change on labour policy and the role of business and organized labour in addressing these impacts, this book finds that climate change has and will have significant impacts on labour markets. Despite this, the consequential requirements for labour market planning are not always priorities for employers' organizations and trade unions. In addition, policy development processes do not always benefit from the important contribution these organizations can bring to the table. The investigations undertaken determined that employers' organizations and trade unions are important actors in ecological modernization. However, if employers' organizations and trade unions are to make a useful contribution to climate change policy, they need to give more consideration to the labour-related impacts of climate change and allocate more of their own resources to make a meaningful contribution to policy in this regard.

This book has also has provided a rationale for the selection of EM as the theoretical framework for investigating policymaking processes and described its validity as a theoretical model for framing policy on climate change. However, and despite the theorists' contentions that EM is shaping the discourse in environmental politics, the international, regional and national policies and policy development processes profiled in this book are not guided by EM or any other theoretical framework. The authors then considered the questions of whether and how EM could be extended as a practical guide to inform the policy development process and, if so, could EM be a tool to bridge the gap that exists between performance and the ambition of those policies? The findings are in the affirmative but within the constraint of the range of, in most cases, the multiple objectives of the policies and the commitments of the stakeholders. As well, EM is still evolving. There are a range of interpretations that can work to blur its usefulness and in some ways the theory is not at a sufficiently mature stage to allow for a definitive analysis of the range of issues that arise during the operational phase – although this book has gone some way towards achieving this.

The process of developing a model to operationalize the theory found it supported three possible outcomes and therefore three possible policy frameworks that are each consistent with EM. It was also found and, as mentioned above, the achievement of the outcome is dependent on policy choices that must also take into account the distinctive political, institutional and cultural features, the national economic importance of the sectors and the extent of the environmental impact on those industries. This is critical in the context of this book, as it is only by committing to an outcome that the state can make the effective policy interventions that will lead to the required outcome. The choice of outcome will dictate whether policy requires a strong science orientation, whether policies will be seeking to deliver progress towards absolute sustainability, or whether they deliver a target such as GHG emission reduction.

Accepting the need for consistency in definition and terminology, the book has presented a standardized platform that can then be used to inform policy interventions and to monitor the progress of implementation. By overlaying an EM template on a sample of national, regional and international climate policies, it was possible to establish which policy initiatives most closely aligned to the 'ideal' EM model. The findings add to the body of knowledge concerning EM and validate its existing theoretical components; however, the findings do need to be updated to reflect current conditions. The investigations, for example, found that EM is not yet sufficiently developed to integrate non-state actors into its theoretical frameworks, neither is it able to specify social partners or other actors beyond the parameters of civil society nor is it sufficiently well-structured to allow for more detailed specification of its theoretical components. Incorporating employers' organizations, trade unions and other civil society actors as structural components in the theory would make EM more consistent with contemporary environmental policy.

The field of climate change policy has evolved rapidly. The negotiations for the 2015 Paris Agreement, the successor treaty to the Kyoto Protocol, were undertaken at a time when the community, industry and governments largely accept the reality of climate change. This is a significant shift in broader government policy and community sentiment from the 2009 Copenhagen COP negotiations, when the real impact of climate change was still the subject of debate. Many stakeholders, including private sector entities, are also now demonstrably contributing on their own initiative, as exemplified by the global retailer Ikea's pledge to clean energy and climate action and shareholder and investor groups' pressure to isolate fossil fuel-dependent activities.

A labour market plan is a key responsibility of the state and industry, both individually and in combination. Individual nations such as France and Germany have written and implemented policies that address the labour market impacts of climate change and suit their specific circumstances while staying in line with a broader international protocol. If the transition to a low carbon economy is to be smooth and effective, the scope of plans should not merely be the single dimension of labour and skill shortages but also the re-crafting of occupational profiles and the emergence of new occupations involved in the delivery and installation of low carbon technologies. To this must be added the effect of new products and regulations, in particular fear of the unknown and asking people to do work for which they have not been properly trained. The green workplace is still a workplace and participants are entitled to be prepared and protected. In regard to workplace policy and strategies, occupational health and safety issues remain an ever-present responsibility of management. In climate change policy, the challenge for the nation-state is to achieve effective policies rather than merely delivering well-intentioned interventions in market behaviour that prove to be ineffective.

There is little doubt that social partners, the employers' organizations and trade unions are active and effective in the climate policy process, but to varying degrees. In the EU, the social partners have an authority conferred by statute, even though they are not necessarily the most capable or most representative organizations. As well and in practice, the contribution of the social partners in the EU member states to the process of developing climate change policy reflects different levels of commitment, a situation common across the organizations in the countries profiled. Their work programme does not always prioritize climate change but rather reflects the current issues of the day.

It was also found that what is a priority issue for the employers' organizations and trade unions in one state may bear no resemblance the work programme in another state that has a different economic profile and where the history, culture and tradition has cultivated different relationships. The UK study in particular demonstrated how culture, history and tradition can frame the statutory influences, attitudes and priorities of an organization. For example, with labour relations, in which the UK government has traditionally adopted a non-interventionist approach, negotiations occur at the level of the workplace and collective bargaining is voluntary. This means that the members' expectations for service by CBI and TUC are different to that from the service delivery expectations of employers' organizations and trade unions members in other countries and where the labour market and the role of the social partners is more regulated.

There is a clear connection between climate change policy and employment and the workplace: a connection that occurs as a consequence of environmental concerns across the community and industry and a commitment through international climate and sustainability agreements to decent work and a just transition. There is a requirement for labour market planning in the strategies and policies associated with combatting climate change and there will be changes to the skills and occupational profiles as a consequence of climate change. Skill shortages could be a barrier to timely adaptation to climate change if they are not addressed in labour market planning.

There is a dichotomy between addressing climate change policy and creating a workforce suitable to take up green jobs. The research on green jobs is well informed but only within its remit, which has largely confined itself to jobs and employment that are directly impacted by climate change, such as work in the renewable energy sector. The contention in this book is that the entire labour market is impacted by climate change and that policies need to address those impacts. In practice, labour market planning and skills policies are not receiving the attention required to ensure the labour is available in the numbers and with the skills required. The OECD observes that while there is agreement about the impact of climate change on employment, there is disagreement about what to do about it. While there is some endorsement for the involvement of employers' organizations and trade unions, the practical demonstrations are limited, providing only vignettes of the contribution to the policy development process.

The peak business organizations and trade unions participate in the intergovernmental proceedings as civil society actors and to that end are advocates alongside many other stakeholders intent on influencing the negotiations in their favour. They are also an important part of the domestic policy development process, providing informed comment and often contributing valuable research to the considerations. Employers' organizations and trade unions act in the context of their respective organizational mandate and in the broader context of lending their influence to the higher order policy outcomes, while also contributing to the broader civil society momentum on particular issues.

When overlaid on the EM template, the UK and EU performed well, arguably on account of the focus of EM being largely on the ecological objectives, whereas the UNFCCC agreements also required policy to complement such issues as poverty eradication, sustainable development and equity. The EU and UK policies are conditional on the achievement of energy security, India on economic development, and China is an economy in transition. Given their constraints, the EU, UK and UNFCCC most closely align to the EM ideal, having many of the required features of technology and innovation, market mechanisms, an important role for the state, engagement with civil society and demonstrating a level of ecological consciousness. In view of the fact that EM can only act as a guide in the policy development process, since those policies must align with the broader social and economic programmes of government, these models can be seen as representing optimum or close to optimum models for EM. The extent to which intergovernmental, regional and national climate-related policymaking processes successfully address stakeholder requirements and ecological imperatives was explored in the country profiles and case studies in some detail.

The country profiles yielded some interesting information. As is seen across the organizations profiled in this study, many of the national organizations rely on the policy papers and research of the peak international organizations to guide their domestic advocacy on climate change and this reliance frames the civil society contribution to the domestic policy and legislative process. Labour market planning and the employment and workplace impacts of climate change have not been a significant issue for these organizations; even the ITUC advocacy addresses only its decent work and just transition mantra. The Green Jobs campaigns initiated by the ILO and supported by UNEP, IOE and ITUC go some way further but maintain the relatively narrow focus of jobs that are considered green, creating a demarcation between the green and not green job but not going further to engage the discussion about the impacts on the labour market that is not yet green. The ICC has not engaged in the green jobs discussions, labour market impacts or commented on the ITUC decent work and just transition advocacy. The IOE mandate is in respect to the interests of business as employers and places a particular focus on the employer interface with the ILO. While the IOE has a clear position on climate change, it is not an active advocate.

In the European Union, the peak representative organizations for business, employers and trade unions are active participants and strong and effective advocates with the European Commission and the European Parliament and in the European Economic and Social Committee, which also has responsibility for climate change and environmental policy. In international negotiations, they have the role as advisors to the EU Party to the COP.

In Germany the peak business association and the employer's organization exist separately with discrete mandates. BDI is the industry organization representing the commercial interests of business and industry in Germany. The peak employer's organization, BDA, has no climate change or environment responsibility.

Australia, Canada and Singapore are developed countries with internationally competitive advanced economies. However, each is vulnerable to the challenges of climate change. Australia, which is the driest inhabited continent on earth and vulnerable to issues such as floods, drought and bushfire, is frequently criticized for the lack of a national climate change policy and action plan. Canada faces the challenges of conflict between developing its diverse energy resources while maintaining its commitment to the environment. Canada withdrew from the Kyoto Protocol in 2011 and pursued the development of its oil tar sands. The change of government in 2015 promises a change in attitude and leadership on climate change and sustainability policy, and to harnessing the authority of its provinces to deliver effective emission reduction programmes. Singapore is a low-lying area that has already experienced rises in sea levels and ambient temperatures. The first Singapore Green Plan was adopted in 1992 and in 2009 it adopted the Sustainable Singapore Green Plan which outlined the sustainable development targets to 2030. The Sustainable Development Blueprint 2015 is the product of feedback obtained from 130,000 people and over the year 2015 6,000 people were engaged through dialogues, surveys and Internet portals.

Canada has no comprehensive federal climate legislation, with authority vested in the regions and states. Also, there are no peak employers' organizations, the authority is with the business associations, employers' organizations and trade unions in the provinces and regions which often act separately from the peak national organizations and make representations to government.

In Singapore, like Germany, the peak employers and business organizations coexist with clearly defined and non-overlapping mandates. While the peak business association, employers' organization and the trade union do not declare an interest in climate change, the parallel economic and environmental goals of the country are reflected in the culture of the organizations.

The Indian government has made a solid commitment to emission reduction but it is constrained by economic, social and environmental factors and to that end, the priority declared in its INDC in advance of the Paris COP was to economic development and poverty eradication ahead of climate change. Employers' organizations and trade unions in India are relatively well resourced as CSOs and as representatives of sections of society, however there are a number of peak organizations and the research for this book selected only the major organizations. The magnitude of the population and land mass and its developing economic situation mitigates against the centralization of representation and accordingly the capacity of the organizations for strategic and effective intervention with the central government. Of the business associations and employers' organizations profiled, none has any stated policy on climate change or any apparent interest in the impacts of climate change on the labour market. The issues confronting many workers in India are basic rights and reasonable reward and these are therefore the focus of the trade union attention.

In Kenya, employers' organizations and trade unions are not engaged in the climate change policy development process. The broader business community also does not engage in this process. This is a reflection of the organizations' focus rather than any apparent barriers to engagement created by the government or departments, which are mandated to engage with the community and stakeholders.

Moving to the case studies, in the case of Europe as a region, investigations for the book found that in EU member states, GHG emission reduction and energy efficiency achievements are defined by the ambition in EU legislation. Additionally, many EU policies assist states in formulating policy and the success of adaptation policies in member states requires that the social dimension is pursued. Looking at the UK specifically, while there has been material commitment and action by the UK government, the work on climate change and climate change policy is far from complete and further significant policy initiatives are necessary to enable the shift to a low carbon economy in the UK. The achievements have been made without a cohesive programme of action by the employers' organizations and trade unions and without particular attention by the government to the labour market issues.

The results of the governance survey were not impressive and presented a range of concerns about the quality of governance of climate policymaking across employers' organizations, trade unions, and other CSOs. In terms of their overall perspectives on climate governance, perhaps the most worrying aspect is that not one sector viewed the implementation capacity of policymaking favourably. On the basis of these results, given that one of the objectives of the Convention (Article 2) is to 'prevent dangerous anthropogenic interference with the climate system' (UN, 1992), it would appear the attitude of these respondents was that the chances of reducing emissions were slim. Here, there is a link to the mixed perspectives about decision-making, which reflect negatively on the ability of climate policymakers to come to agreement about the best way forward. The ratings from employers demonstrated that policymaking lacked transparency and accountability, and from unions that equality was absent. Both of these results indicate a lack of confidence in certain aspects around the governance of climate policymaking. The most universally negative perception concerned the level of resources provided to stakeholders to represent their interests. Even if the venues of policymaking acted with more responsibility and improved decision-making and implementation arrangements, business and labour would still have to fund their own participation if they wanted to get a seat at the table. While the Paris Agreement contains a range of provisions about capacity building (Article 11), transparency (Article 13) and implementation (Article 15), it will take some time to see if these make any difference for these participants.

This book has brought to light a number of issues that invite further consideration. The research contention relied on the view that employers' organizations and trade unions were active in labour market planning and that, if climate change impacted the labour market, they would have a role in the development of policy in response. The contention was found to be incorrect and employers' organizations and trade unions did not consistently prioritize the labour market dimension of climate change policy and did not effectively participate in labour market planning. Although the contention was found to be incorrect, there remains the expectation by agencies such as the OECD, the EU and the ILO that there is a need for labour market planning in climate change and that employers' organizations and trade unions are the appropriate bodies to do this work. To that end, it is recommended further research be undertaken to determine the conditions required for employers' organizations and trade unions to prioritize labour market planning and to engage effectively in the process.

The book has also extended the theoretical framework of ecological modernization to an operationalized model. It established that there can be more than one pathway and policy framework. To that end, it is recommended that further research be undertaken to first, harmonize the multiple interpretations, and then to develop the concept of multiple pathways that present as options in the process of its operationalization. It was also found that the opportunities for employers' organizations and trade unions as civil society actors were enhanced in the stronger versions of EM, and that in international climate change negotiations they were influential actors in the civil society movement. Further research that develops the role of employers' organizations and trade unions as civil society actors would potentially advance the understanding of the role of civil society in the operationalizing of EM.

Consequently, participation of CSOs in climate change-related deliberations, particularly around issues of inclusiveness, equality, transparency and resources, must be addressed both by ecological modernization theory and policymaking processes if business and labour are truly to have a seat at the negotiating table and make a productive contribution to the desired ecological outcome.

Appendix: research materials for participants

INTRODUCTORY LETTER

Date

Dear . . .

Research Project: Business, Organized Labour and Climate Policy: Forging a Role at the Negotiating Table

It would be greatly appreciated if you would contribute information about your organization for this research project. The research is being undertaken by Dr Tim Cadman of Griffith University and will explore the role of employers' organizations and trade unions in the development of climate change policy. As in important actor in the process your organization's experience will be valuable and will help shape the research findings which are to be published in a book with the working title as above.

The findings will be drawn from a qualitative study of the role of the organizations in approximately 10 countries. The underlying context is that the labour market will be significantly disrupted by the changed patterns of production and consumption that are essential in effective climate change policy, and as the representatives of the stakeholders in the labour market employers organizations and trade unions will be advocates in the process.

Details of the research project scope and aims, and the researchers are attached. Also attached is the organizational data template that we ask you to complete and return by the end of February. We are not asking for confidential information and while some of the information sought is publicly available, your contribution will ensure the research completeness of the data collection. A second phase of the data collection will be conducted during March and April and will enquire of your perceptions about the process, your role and the effectiveness of your engagement.

This is an interesting and important field of study and one on which very little academic research has been undertaken. The research project will contribute to the body of knowledge and we welcome your contribution.

If you would like to discuss the project or would like further information, please let me know.

Regards
Dr Tim Cadman

Attachments:
Consent form
Participant information sheet
Organizational data template

CONSENT FORM

Business, Organized Labour and Climate Policy: Forging a Role at the Negotiating Table

Principal Researcher: Dr Tim Cadman

The conduct of this research involves the collection, access and/or use of your identified personal information. The information collected is confidential and will not be disclosed to third parties without your consent, except to meet government, legal or other regulatory authority requirements. A de-identified copy of this data may be used for other research purposes. However, your anonymity will at all times be safeguarded. For further information, consult the University's Privacy Plan at http://www.griffith.edu.au/privacy-plan or telephone (07) 3735 4375.

- I have read the Participant Information Sheet and the nature and purpose of the research project has been explained to me.
- I understand and agree to take part.
- I understand the purpose of the research project and my involvement in it.
- I understand that I may withdraw from the research project at any stage and that this will not affect my status now or in the future.
- I confirm that I am over 18 years of age. *Omit if participants are under age of 18.*
- I understand that while information gained during the study may be published, I will not be identified and my personal results will remain confidential. *If other arrangements have been agreed in relation to identification of research participants this point will require amendment to accurately reflect those arrangements.*

- I understand that the interview and/or workshop in which I will participate will be recorded.
- I understand that the tape will be retained for five years in digital format located only on the personal computer of Dr Tim Cadman, after which it will be destroyed.

❒ I consent to this INTERVIEW being recorded.

Name of participant. . . . .
Signed. . . . . .Date. . . . .

Griffith University conducts research in accordance with the National Statement on Ethical Conduct in Human Research. If potential participants have any concerns or complaints about the ethical conduct of the research project they should contact the Manager, Research Ethics 07 373 54375 or research-ethics@griffith.edu.au

PARTICIPANT INFORMATION SHEET

Business, Organized Labour and Climate Policy: Forging a Role at the Negotiating Table

TO: All civil society stakeholders in climate policy negotiations (especially business and labour)

From: Dr Tim Cadman, Griffith University, Dr Peter Glynn research assistant

Re: Invitation to participate in the research project: Forging a seat at the negotiating table: business and labour in climate policy negotiations

DATE:

Dear Stakeholder,
We would like to invite you to take part in this research project. We are conducting research into the role of civil society in climate policy negotiations, especially business and labour organizations and stakeholder attitudes in particular. We are looking at climate policy negotiations at various levels:

- At the intergovernmental level (e.g. within the climate talks under the United Nations Framework Convention on Climate Change – UNFCCC);

- At the regional level (e.g. within the European Union);
- At the national level (e.g. within the United Kingdom);
- At other levels (e.g. at the municipal level).

Please read the information provided below carefully. Its purpose is to explain to you as openly and clearly as possible all the procedures involved so that you can make a fully informed decision as to whether you are going to participate. Feel free to ask questions about any information in the materials. You may also wish to discuss the project with a relative or friend or your local health worker. Feel free to do this. You are invited to participate once you understand what the project is about, but you may withdraw from the project at any time.

Who is conducting the research?
The principal researcher is Dr Tim Cadman, Institute for Ethics Governance and Law, Griffith University, Nathan, Queensland 4111 Australia; Ph: +61 419 628709; email: t.cadman@griffith.edu.au. He is being assisted by Dr Peter Glynn in a private research capacity.

Why is the research being conducted?
The research focuses on civil society engagement in the development of climate change policy, and in particular the role of employers' organizations and trade unions. The impacts of climate change policy will be felt in all workplaces and across all sectors of industry. Those impacts must be reflected in plans for the labour market if business is to continue to grow and if the transition to a low carbon economy is to be fair for workers.

What you will be asked to do?
Participation in this project will involve:

- An Internet-based questionnaire asking stakeholders, who play a part in sustainable development what they think of their involvement in the initiatives in question and related activities.
- There are 18 questions, and opportunities for comments. The survey can take as little as ten minutes; making comments will last longer, depending on how much the participant wishes to write.
    - The survey invites you to be involved in a follow-up interview; space is provided for people to provide their contact details. Agreement to participate in follow-up interviews is entirely voluntary. You will be interviewed for no more than 50 minutes at a place and date of mutual convenience.
        - Please note that the interviews will be recorded, and this part of the research will require you to provide your personal

information. This information will be kept confidential, and used by the researchers only.

- You may add further comments if you wish.
- You may withdraw from the survey at any time. However, by clicking 'done', you indicate that you understand the information and that you give your consent to participate in the research project. Your participation in the survey is anonymous.

The expected benefits of the research
The research will help people who play a part in climate policy negotiations to decide if they think these negotiations are being run and managed in the proper way. This will help them decide if they want to keep playing a role, or if it is better not to participate. The intention of the project is to provide research findings, to improve the climate policy negotiations themselves for the benefit of participants, and climate policy generally.

Risks to you
There are minimal risks with this survey and the interview. It is possible you may feel a sense of frustration with the climate policy negotiations as a result of writing about, or sharing, your feelings.

Your confidentiality
Although the survey is anonymous, the conduct of this research involves the collection, access and/or use of your identified personal information should you wish to be interviewed. The information collected is confidential and will not be disclosed to third parties without your consent, except to meet government, legal or other regulatory authority requirements. A de-identified copy of this data may be used for other research purposes. However, your anonymity will at all times be safeguarded. For further information consult the University's Privacy Plan at http://www.griffith.edu.au/privacy-plan or telephone (07) 3735 4375.

Your participation is voluntary
Participation is entirely voluntary. If you do not wish to take part you are not obliged to. If you decide to take part and later change your mind, you are free to withdraw from the project at any stage. Any information already obtained from you will be destroyed. Your decision whether to take part or not to take part, or to take part and then withdraw, will not affect your relationship with Griffith University.

Questions / further information
Further information about the project is available at:
https://www.dropbox.com/s/2l87hrk8k7ehlmu/Research project overview.docx?dl=0

Should you have any queries regarding the progress or conduct of this research, you can contact the principal researcher:
Doctor Tim Cadman

Research Fellow
Institute for Ethics Governance and Law
Griffith University
Nathan QLD 4111
Ph: 0419 628 709
Email: t.cadman@griffith.edu.au

The ethical conduct of this research
Griffith University conducts research in accordance with the National Statement on Ethical Conduct in Human Research. If potential participants have any concerns or complaints about the ethical conduct of the research project they should contact the Manager, Research Ethics 07 373 54375 or research-ethics@griffith.edu.au.

ORGANIZATIONAL DATA TEMPLATE

Instructions: This is a word document and the table will expand to take all the information entered. Please provide as much information as is available and necessary to answer each section. Also include as reference source of the information including web links, full document titles and page or paragraph numbers.

Organization:

Name:	
Legal status:	
Address and contact information:	
Membership: numbers and type:	
Jurisdiction: geographic, demographic, sectors:	
Organization aims and objective:	
Structure: staff number, departments:	

External representations:	
Internal committees:	
Member engagement:	
Responsible officer re climate change:	

Climate change credentials:

Organization climate change or sustainability policy:	
Strategy to achieve policy objectives:	
Resources available: budget, staff, committees:	
Publications, commissioned research:	
Lobbying and communication strategy:	
Member engagement:	
Representation on government and other peak regional and international bodies:	
Membership of consortia/ lobby groups:	
KPIs, monitoring and evaluation:	

References

ACEI (Alliance for a Competitive European Industry). (2013). Alliance for a competitive European industry. Retrieved on 7 November 2016 from www.businesseurope.eu/content/default.asp?PageID=605.

ADEME (L'Agence de l'Environnement et de la Maîtrise de l'Énergie). (2011). The social housing stock. Retrieved on 7 November 2016 from www.plan-batiment.legrenelle-environnement.fr/index.php/g-les-4-secteurs/le-parc-des-logements-sociaux.

Ashford, N.A. (2002). Governmental and environmental innovation in Europe and North America. *American Behavioural Scientist*, 45(9), 1417–1434.

Australian Legal Dictionary. (2015). Definition: Non-state actors. Retrieved on 7 November 2016 from http://bond.libguides.com/dictionaries-encyclopaedia-commentaries/dictionaries.

Bailey, I. and Rupp, S. (2006). The evolving role of trade associations in negotiated environmental agreements: The case of United Kingdom climate change agreements. *Business Strategy and the Environment*, 15(1), 40–54.

Bailey, I., Gouldson, A. and Newell, P. (2011). Ecological modernisation and the governance of carbon: A critical analysis. *Antipode*, 43(3), 682–703.

Barker, A. and Clark, P. (2015). Indian stance slows progress towards climate change deal. *Financial Times*, 17 November, p. 6.

Barnard, C. (2002). The social partners and the governance agenda. *European Law Journal*, 8(1), 80–101.

BBC. (2016). Country profiles. Retrieved on 8 November 2016 from http://news.bbc.co.uk/2/hi/country_profiles/default.stm.

BDI. (2016a). Resource efficiency in the circular economy. Berlin: Bundesverband der Deutschen Industrie e.V.

BDI. (2016b). Industry supports circular economy initiative. Berlin: Bundesverband der Deutschen Industrie e.V.

BDI. (2016c). Legal and planning certainty for operators of industrial installations. Berlin: Bundesverband der Deutschen Industrie e.V.

BDI. (2016d). Improvement proposals for the BREF process. Berlin: Bundesverband der Deutschen Industrie e.V.

BDI. (2016e). Entrepreneurial freedom vs. further development of medium-related environmental protection. Berlin: Bundesverband der Deutschen Industrie e.V.

BDI. (2016f). More than 300 different soil types in Europe make one-size-fits-all soil protection legislation difficult. Berlin: Bundesverband der Deutschen Industrie e.V.

BDI. (2016g). Bring economy and ecology into line with each other. Berlin: Bundesverband der Deutschen Industrie e.V.

Beck, U. (1992). *Risk society. Towards a new modernity*. London: Sage Publications.

BIS. (2008). *Lisbon strategy for jobs and growth: UK national reform programme*. London: Department for Business, Industry and Skills.

BIS. (2012). *Trade union membership 2011*. London: Department for Business, Industry and Skills, UK.

BIS. (2016). *Trade union membership 2015: Statistical bulletin*. London: Department for Business, Innovation and Skills.

Blackstock, K.L., Richards, C., Kirk, E.A., Chang, Y.C. and Davidson, G. (2006). Participation and regulation: Where two worlds collide? *Participatory Approaches in Science and Technology Conference, Edinburgh*. Retrieved on 7 November 2016 from www.Macaulay.Ac.uk/PATHconference/outputs/PATH_abstract_1.1.1.Pdf.

Blackwell. (1999a). *The Blackwell dictionary of political science*: Employers organisations. Retrieved on 7 November 2016 from www.credoreference.com.ezproxy.bond.edu.au/entry/bkpolsci/employers_organisations.

Blackwell. (1999b). *The Blackwell dictionary of political science*: Trade unions. Retrieved on 7 November 2016 from www.credoreference.com.ezproxy.bond.edu.au/entry/bkpolsci/trade_unions.

BPIE. (2011). *Europe's buildings under the microscope: A country-by-country review of the energy performance of buildings*. Brussels: Buildings Performance Institute Europe.

Branger, F., Lecuyer, O. and Quirion, P. (2015). The European Union Emissions Trading Scheme: should we throw the flagship out with the bathwater? *Wiley Interdisciplinary Reviews: Climate Change*, 6(1), 9–16.

Braun, R. (2010). Social participation and climate change. *Science and Business Media*, 12, 777–806.

Brazil CSD. (2012). *Rio + 20 guide*. Rio de Janeiro: UNCSD.

Burrow, S. (2015). How is the global organised labour movement responding to the threat of climate change? Retrieved on 7 November 2016 from www.weforum.org/agenda/2015/12/how-will-climate-change-affect-jobs/.

Business and Climate Summit. (2015). *Business and climate summit conclusions: Towards a low-carbon society*. Paris: Business and Climate Summit.

BusinessEurope. (2012). *Doha climate conference: Absence of level playing field requires EU climate and energy rethink*. Brussels: BusinessEurope.

BusinessEurope. (2013a). A competitive EU energy and climate policy. BusinessEurope recommendations for a 2030 framework for energy and climate policies. Brussels: BusinessEurope.

BusinessEurope. (2013b). Climate change. Retrieved on 8 November 2016 from www.businesseurope.eu/Content/Default.asp?PageID=657.

BusinessEurope. (2013c). Mission and priorities. Retrieved on 8 November 2016 from www.businesseurope.eu/content/default.asp?PageID=582.

BusinessEurope. (2013d). *Debate on the EU 2030 energy and climate framework* (14 February). Brussels: BusinessEurope.

BusinessEurope. (2013e). Employment and social affairs. Retrieved on 8 November 2016 from www.businesseurope.eu/content/default.asp?PageID=592.

BusinessEurope. (2013f). BusinessEurope urges the 22 May 2013 European Council to focus on lifting obstacles to growth. *BusinessEurope Press Release*, 17 May, p. 1.

BusinessEurope. (2015a). On the road to Paris snapshot of energy efficiency technologies. Brussels: BusinessEurope.

BusinessEurope. (2015b). On the road to Paris. A global deal is our business (June). Brussels: BusinessEurope.

BusinessEurope. (2016). BusinessEurope comments on the Commission's EU ETS reform proposal. Brussels: BusinessEurope.

Buttel, F.H. (2000). Ecological modernisation and social theory. *Geoforum*, 31, 57–65.

C20. (2012). Civil20 process. Retrieved on 7 November 2016 from http://dialogues.civil20.org/G20_process.

Cabré, M.M. (2011). Issue-linkages to climate change measured through NGO participation in the UNFCCC. *Global Environmental Politics*, 11(3), 10–22.

Cadman, T. (2011). *Quality and legitimacy of global governance: Case lessons from forestry*. New York; Basingstoke: Palgrave Macmillan.

Cadman, T. (2015). Lost shoes and teargas highlight mixed messages from Paris, 29 November. Retrieved on 7 November 2016 from timcadman.wordpress.com.

Cadman, T., Maraseni, T., Ma, H.O. and Lopez-Casero, F. (2016). Five years of REDD+ governance: The use of market mechanisms as a response to anthropogenic climate change. *Forest Policy and Economics*. doi:10.1016/j.forpol.2016.03.008.

Cambridge Dictionary. (2013). Definition: Employer organisation. Retrieved on 7 November 2016 from http://dictionary.cambridge.org/dictionary/business-english.

Cameron, E., Erickson, C., Prattico, E. and Schuchard, R. (2015). *Creating an action agenda for private-sector leadership on climate change (BSR Working Paper: Business in a climate-constrained world)* (2nd edn). San Francisco, CA: BSR.
Carbon Pulse. (2016a). Analysts say Australia can meet CO_2 targets with current policies, but market needed. Retrieved on 8 November 2016 from http://carbon-pulse.com/19439/.
Carbon Pulse. (2016b). CP daily 17 May 2016. Retrieved on 7 November 2016 from http://carbon-pulse.com/19829/.
Carbon Pulse. (2016c). Germany launches €17 billion campaign to boost energy efficiency. Retrieved on 7 November 2016 from http://carbon-pulse.com/19829/.
CARE Australia. (2010). Community based adaptation enabling environment. NCCARF Conference, June 2010, Gold Coast, Poster Presentation.
CBI. (2007). *Climate change: Everyone's business*. London: Confederation of British Industry Climate Change Task Force.
CBI. (2009). *All together now: A common business approach for greenhouse gas emissions reporting*. London: Confederation of British Industry.
CBI. (2011). *Mapping the route to growth, rebalancing employment* (Brief, June 2011). London: Confederation of British Industry.
CBI. (2012a). *Climate change and business: The role of business*. Retrieved on 21 November 2016 from www.cbi.org.uk/business-issues/energy-and-climate-change/climate-change-and-business/the-role-of-business.
CBI. (2012b). *The colour of growth. Maximising the potential of green business*. London: Confederation of British Industry.
CBI. (2014a). *A climate for growth: Securing a global climate change deal in Paris*. London: Confederation of British Industry.
CBI. (2014b). *Business and public attitudes towards UK energy priorities*. Retrieved on 21 November 2016 from news.cbi.org.uk/business. . ./energy. . ./publications/business-and-public-attitudes-towards-UK-energy-priorities.
CBI. (2015a). *Setting the bar: Energy and climate change priorities for the government*. London: Confederation of British Industry.
CBI. (2015b). *Priorities for the new government*. London: Confederation of British Industry.
CBI. (2015c). *Effective policy, efficient homes refreshing the UK's approach to retrofitting homes*. London: Confederation of British Industry.
CBI. (2015d). *Small steps, big impact – maximising the role of on-site generation in meeting our energy and climate change challenges*. London: Confederation of British Industry.
CCC. (2013). *$50 million a day*. Montreal: Canadian Chamber of Commerce.
CCC. (2014). The measures that matter: How Canada's natural resource sector is working to protect the environment. Retrieved on 21 November

2016 from www.chamber.ca/media/news-releases/141223-the-measures-that-matter/.

CCC. (2015). What does COP21 mean for Canadian business? Retrieved on 21 November 2016 from http://bit.ly/1Ze2o3K.

CEDEFOP. (2010). *Skills for green jobs, European synthesis report*. Luxembourg: European Union.

Chan, S. (2015). Rich nations are urged to pay climate change tab. *International New York Times*, 5 December, p. 5.

Christ, R. (2014). *Statement of Renate Christ, secretary of the IPCC at the opening of SBSTA-40 Bonn*, 4 June. Bonn, Germany: Intergovernmental Panel on Climate Change.

Christoff, P. (1996). Ecological modernisation, ecological modernities. *Environmental Politics*, 5(3), 476–500.

Christoff, P. (2016). Ideas for Australia: A six-point plan for getting climate policy back on track. *The Conversation*, 14 April.

CIA. (2016). CIA fact book. Retrieved on 8 November 2016 from www.ciaworldfactbook.us/I.html.

Clark, P. and Stothard, M. (2015). India and France seek to raise $1tn for cheap solar power as climate talks open. *Financial Times*, 1 December, p. 1.

CLC. (2015a). Collaborative approach will be key to realizing Canada's climate change obligations. Retrieved on 12 December 2015 from http://canadianlabour.ca/news/news-archive/collaborative-approach-to-climate-change.

CLC. (2015b). *ENERGY – alternatives for a green economy*. Montreal: Canadian Labour Congress.

CLC. (2015c). Labour delegation to champion a just transition to a green economy at Paris climate change summit. Retrieved on 26 November 2015 from http://canadianlabour.ca/news/news-archive/labour-delegation-champion-just-transition-green-economy-paris-climate-change.

CLC. (2015d). *Making the shift to a green economy: A common platform of the green economy network*. Montreal: Canadian Labour Congress.

CLC. (2015e). *Reducing greenhouse gas emissions in Canada*. Montreal: Canadian Labour Congress.

CLC. (2016a). David suzuki joins with CLC to support a one million climate jobs plan' (2016). Retrieved on 2 March 2016 from http://canadianlabour.ca/news/news-archive/david-suzuki-joins-clc-support-one-million-climate-jobs-plan.

CLC. (2016b). *One million climate jobs. A challenge for Canada*. Montreal: Canadian Labour Congress.

Climate Action Network. (2012). ECO 7, RIO + 20, English version. *CAN Newsletter*, *Rio + 20*, retrieved on 4 November 2016 from www.climatenetwork.org/newsletter/eco-7-rio20-english-version.

Cohen, M. (1997). Risk society and ecological modernisation. Alternative visions for post-industrial nations. *Futures*, 29(2), 105–119.

Cohen, M. (2000). Ecological modernisation, environmental knowledge: A preliminary analysis of the Netherlands. *Environmental Politics*, 9(1), 77–106.

Combet, G. (2010). *Ministerial statement; home insulation program*. Canberra: Department of Climate Change and Energy Efficiency, Commonwealth of Australia.

COP/CMP Peruvian Presidency, French Presidency, UNFCCC LPAA. Vision and approach. Paris: Paris 2015 UN Climate Change Conference COP 21-CMP 11.

Council of Europe. (1996). European Social Charter (Revised), 3 May 1996, ETS 163.

Cridland, J. (2011). *Achieving competitive, secure energy in a low-carbon economy*. London: Confederation of British Industry.

Cridland, J. (2012). The heat is on. We need decisions on energy. *The Times*, 19 April.

Croucher, R., Tyson, S. and Wild, A. (2006). 'Peak' employers' organisations: International attempts at transferring experience. *Economic and Industrial Democracy*, 27, 463–484.

Dalber Global Development Advisors. (2009). *Climate change, the anatomy of a silent crisis: Human impact report*. Geneva: Global Humanitarian Forum.

Davenport, C. and Harris, G. (2015). 'Future of life' at stake in talks. *International New York Times*, 1 December, pp. 1–4.

de Boer, R., Benedictus, H. and van der Meer, M. (2005). Broadening without intensification: The added value of the European social and sectoral dialogue. *European Journal of Industrial Relations*, 11(1), 51–70.

DECC. (2007). *Meeting the energy challenge: A White Paper on energy*. London: Department of Energy and Climate Change.

DECC. (2009). *The UK low carbon transition plan*. (White Paper, Amended 20 July 2009 from the version laid before Parliament on 15 July 2009). London: Department of Energy and Climate Change.

DECC. (2010). *2050 pathways analysis*. London: Department of Energy and Climate Change.

DECC. (2011a). *The carbon plan: Delivering a low carbon future*. London: Department of Energy and Climate Change.

DECC. (2011b). *Climate change act 2008* (No. 2008 Chapter 27). London: Department of Energy and Climate Change.

DECC. (2011c). The greenhouse gas effort sharing decision. Retrieved on 8 November 2016 from www.decc.gov.uk/en/content/cms/emissions/ghgsd/ghgsd.aspx.

DECC. (2011d). *Energy Act 2011*. London: Department of Energy and Climate Change.
DECC. (2011e). *3rd progress report*. London: Department of Energy and Climate Change, Committee on Climate Change.
DECC. (2011f). *Progress report 2011: Adapting to climate change in the UK. Measuring progress*. London: Department of Energy and Climate Change, Committee on Climate Change, Adaptation Sub Committee.
DECC. (2012a). 2050 pathways analysis, the 2050 challenge. Retrieved on 8 November 2016 from www.decc.gov.uk/en/content/cms/tackling/2050/3050.aspx.
DECC. (2012b). Davey puts energy saving at heart of strategy. 2012 (09), 8 February, www.decc.gov.uk/en/content/cms/news/pn12_009/pn12_009.aspx.
DECC. (2012c). Green deal. Retrieved on 8 November 2016 from www.decc.gov.uk/en/content/cms/tackling/green_deal/green_deal.aspx.
DECC. (2012d). The green deal and energy company obligation consultation. Retrieved on 8 November 2016 from www.decc.gov.uk/en/content/cms/consultations/green_deal/green_deal.aspx.
DECC. (2014). *Meeting carbon budgets-2014 progress report to parliament*. London: Depaerment of Energy and Climate Change.
DECC. (2015). *Reducing emissions and preparing for climate change: 2015 progress report to parliament*. London: Department of Energy and Climate Change.
Dresner, S., Jackson, T. and Gilbert, N. (2006). History and social responses to environmental tax reform in the United Kingdom. *Energy Policy*, 34(8), 930–939.
Dryzek, J.S., Downes, D., Hunold, C., Schlosberg, D. and Hernes, H. (eds) (2009). *Ecological modernisation, risk society, and the green state*. Oxon: Routledge.
Eastwood, L. (2011). Climate change negotiations and civil society participation: Shifting and contested terrain. *Theory in Action*, 4 (1), 8–38.
Ebbinghaus, B. (2002). Trade unions' changing role: Membership erosion, organisational reform, and social partnership in Europe. *Industrial Relations Journal*, 33(5), 465–483.
EC (European Commission). (2006). *Action plan for energy efficiency: Realising the potential* (No. COM (2006) 545 final). Brussels: European Commission.
EC (European Commission). (2011a). *Consulting European social partners: Understanding how it works*. Luxembourg: European Union.
EC (European Commission). (2011b). *Industrial relations in Europe 2010*. Luxembourg: European Commission.
EC (European Commission). (2011c). *White Paper: Roadmap to a single*

European transport area-towards a competitive and resource efficient transport system. Brussels: European Commission.

EC (European Commission). (2013). *Industrial relations in Europe 2012*. Luxembourg: European Commission, Directorate-General for Employm ent, Social Affairs and Inclusion.

EC (European Commission). (2014). *2030 climate and energy framework*. Brussels: European Commission.

EC (European Commission). (2015). *Industrial relations in Europe 2014*. Luxembourg: European Commission, Directorate-General for Employment, Social Affairs and Inclusion.

ECA UK. (2011). *2021 vision: The future of the electrical contracting industry*. London: NICEIC, Electrical Contractors Association UK.

ECO. (2015). Transparency: Just a castle on a cloud? *ECO*, (7), 2.

Environment Canada. (2007). *Action on climate change and air pollution*. Toronto: Environment Canada.

Environment Canada. (2011). *Federal adaptation policy framework*. Toronto: Government of Canada.

Environment Canada. (2016). *Canada's way forward on climate change*. Toronto: Government of Canada.

Esty, D. and Porter, M.E. (2005). National environmental performance: An empirical analysis of policy results and determinants. *Environment and Development Economics*, 10(430), 391–434.

ETUC. (2004). Union proposal for a European policy on climate change. Retrieved on 8 November 2016 from www.etuc/a/250.

ETUC. (2007). *Climate change and employment. Impact on employment in the EU-25 of climate change and CO_2 emission reduction measures by 2030*. Brussels: European Trade Union Confederation.

ETUC. (2010). *Climate change: The new industrial policies and ways out of the crisis*. Brussels: European Trade Union Confederation.

ETUC. (2011a). *European social partners dialogue on climate change*. Brussels: European Trade Union Confederation.

ETUC. (2011b). *Resolution: European Commission's transport White Paper* (No. Executive Committee Meeting 19–20 October 2011). Brussels: European Trade Union Confederation.

ETUC. (2012). Rio + 20: Social justice can only be guaranteed with environmental protection. Retrieved on 21 November 2016 from www.etuc.org.

ETUC. (2013a). *A new path for Europe: ETUC plan for investment, sustainable growth and quality jobs*. Brussels: European Trade Union Confederation.

ETUC. (2013b). Structure. Retrieved on 8 November 2016 from www.etuc.org/r/11.

ETUC. (2013c). Our aims. Retrieved on 8 November 2016 from www.etuc.org/r/2.

ETUC. (2013d). *Holistic transport management and trade unions*. Brussels: European Trade Union Confederation.

ETUC. (2015a). ETUC position on the structural reform of the EU emissions trading system. Brussels: European Trade Union Confederation.

ETUC. (2015b). European business, local authorities, civil society and trade unions want EU leaders to live up to their Paris commitments. Brussels: European Trade Union Confederation.

ETUC. (2015c). ETUC key demands for the climate COP21 (No. Executive Committee of 17–18 June 2015). Brussels: European Trade Union Confederation.

ETUC. (2016). *ETUC declaration on the Paris Agreement on climate change*, 15 January. Brussels: European Trade Union Confederation.

ETUI. (2011). *Benchmarking working Europe 2011* (No. 13/2011/10574/08). Brussels: The European Trade Union Institute.

Eupolitix. (2013). European Union law. Retrieved on 7 November 2016 from www.eupolitix.com/eu-law.html.

Euractiv. (2010). France catching up with other member states on environment. Retrieved on 7 November 2016 from www.euractiv.com.

Euractiv. (2012). Brussels rolls out carbon market fix. Retrieved on 8 November 2016 from www.euractiv.com/print/climate-environment/.

Euractiv. (2013). EU summit set to turn agenda on climate change upside down. Retrieved on 8 November 2016 from www.euractive.com/energy.

Eurofound. (2009). *Greening the European economy: Responses and initiatives by member states and social partners* (No. EF/09/72/EN). Dublin: European Foundation for the Improvement of Living and Working Conditions.

Eurofound. (2011a). *Industrial relations and sustainability: The role of social partners in the transition to a green economy* (EF//11/26/EN). Dublin, Republic of Ireland: European Foundation for the Improvement of Living and Working Conditions.

Eurofound. (2011b). Community charter of the fundamental social rights of workers. Retrieved on 8 November 2016 from www.eurofound.europa.eu/areas/industrialrelations/dictionary/definitions/communitycharteroft hefundamentalsocialrightsofworkers.htm.

Eurofound. (2011c). European social partners. Retrieved on 7 November 2016 from www.eurofound.europa.eu/areas/industrialrelations/diction ary/definitions/europeansocialpartners.htm.

Eurofound. (2011d). *Industrial relations and sustainability: The role of social partners in the transition to a green economy* (No. EF//11/26/EN).

Dublin: European Foundation for the Improvement of Living and Working Conditions.

Eurofound. (2015). *Collective bargaining in Europe in the 21st century*. Dublin, Republic of Ireland: European Foundation for the Improvement of Living and Working Conditions. doi:10.2806/210184.

Europa. (2010a). Environment and employment. More than the sum of their parts: The links between environment and employment policies. Retrieved on 8 November 2016 from http://ec.europa.eu/environment/integration/employment_en.htm.

Europa. (2010b). The EU climate and energy package. Retrieved on 8 November 2016 from http://ec.europa.eu/environment/climat/climate_action.htm.

Europa. (2012a). Environmental integration. Retrieved on 4 November 2016 from ec.europa.eu/environment/integration/integration.htm.

Europa. (2012b). Treaty of Maastricht on European Union. Retrieved on 7 November 2016 from http://europa.eu/legislation_summaries/institutional_affairs/treaties_maastricht.

Europa. (2012c). Treaty on European Union (Consolidated Version), Treaty of Maastricht, 7 February 1992, Official Journal of the European Communities C 325/5.

Europa. (2016). EU 2030 climate and energy framework. Retrieved on 8 November 2016 from http://ec.europa.eu/clima/policies/strategies/2030/index_en.htm.

European Community. (1999). Treaty Establishing the European Community (1997) O.J. (C 340), 173–308.

European Parliament. (2001). Charter of the fundamental rights of the European union. Retrieved on 8 November 2016 from www.europarl.europa.eu/charter/default_en.htm.

European Parliament. (2007). Charter of fundamental rights of the European union, Treaty U.S.C. (2007).

European Union. (2010). *Consolidated versions of the treaty on European union and the treaty on the functioning of the European union 2010/C 83/01*. Brussels: European Union.

Fabius, L. (2015). Discours de M. Laurent Fabius, Ministre des Affaires Étrangères et du Développement international [speech made by Mr. Laurent Fabius, Minister of Foreign Affairs and International Development]. Speech made at Opening Plenary of the 42nd Session of the UNFCCC Subsidiary Body on Scientific and Technical Advice.

Falkner, G. (2011). The council or the social partners? EC social policy between diplomacy and collective bargaining. *Journal of European Public Policy*, 7(5), 705–724.

Fisher, D. (2010). COP-15 in Copenhagen: How the merging of movements left civil society out in the cold. *Global Environmental Politics*, 10(2), 11–17.

FoE (Friends of the Earth). (2008). History of the Big Ask. Retrieved on 7 November 2016 from www.foe.co.uk/news/big_ask_history_15798.html.

Food and Beverage Leaders. (2015). Dear US and global leaders: This could be the turning point. *Financial Times* (1 October), 5.

Friedl-Schafferhans, M. (2011). *Qualification-green jobs*. Brussels: Walqing-European Trade Union Institute.

Frontier Economics. (2013). Lessons learnt from the current energy and climate framework A report prepared for BusinessEurope, May 2013. Brussels: BusinessEurope.

Garman, J.D.R. and Pearson, P. (2015). 'Green collar nation: A just transition to a low-carbon economy'. London: Trade Union Congress.

Garnaut, R. (2008). *The Garnaut climate change review*. Melbourne: Cambridge University Press.

Garnaut, R. (2011). *The Garnaut climate change review; update 2011: Carbon pricing and reducing emissions* (Update Paper No. 6). Canberra: Australian Government.

GGKP. (2013). *Scoping paper: Moving towards a common approach on green growth indicators*. Geneva: Green Growth Knowledge Platform.

GHK Consulting. (2007). *Links between the environment, economy and jobs* (No. J3476 – Final report). London: European Commission, DG Environment.

Gillis, J. and St. Fleur, N. (2015). Major companies join climate change efforts. *International New York Times* (24 September), 15.

Gleeson, C., Yang, J. and Lloyd-Jones, T. (2011). *European retrofit network: Retrofitting evaluation methodology report*. London: University of Westminster.

Glynn, P.J. (2014). *The role of employers organisations and trade unions in the development of climate change policy: A European perspective* (PhD dissertation).

GOI. (2008). *National action plan on climate change*. New Delhi: Government of India.

GOI. (2015). India climate portal. Retrieved on 8 November 2016 from http://indiaclimateportal.org/the-napcc/.

Golmohammadi, B. (2012). Civil society and Rio + 20. *UN Chronicle*, 49(1/2), 7–8.

Gonzales, G.A. (ed.) (2009). *Democratic ethics and ecological modernisation*. Oxon: Routledge.

Goyal, M. and Sharma, N. (2012). Why India industry lobby groups like CII, FICCI and Nasscom have lost sheen. *The Economic Times*, 3 June.

Guyatt, G., Oxman, A., Vist, G., Kunz, R., Falck-Ytter, Y., Alonso-Coello,

P. and Schünemann, H. (2008). Grade: An emerging consensus on rating quality of evidence and strength of recommendations. *British Medical Journal*, 336 (April), 924–926.

Hache, A. (2011). *Under the radar: The contribution of civil society and third sector organisations to inclusion*. Luxembourg: European Commission Joint Research Centre.

Hajer, M. (1995). *The politics of environmental discourse: Ecological modernization and the policy process*. London: Oxford University Press.

Hall, N.L. and Taplin, R. (2007). Solar festivals and climate bills: Comparing NGO climate change campaigns in the UK and Australia. *Global Warming: Energy Security or Energy Sovereignty?* 1–15.

Hampton, P. (2015). *Workers and trade unions for climate solidarity. Tackling climate change in a neoliberal world*. London: Routledge.

Hayden, A. (2014). *Climate change, ecological modernisation, and sufficiency. When green growth is not enough*. Montreal: McGill – Queens University Press.

Hayes, M. (2007). *Incrementalism*. London: Sage Publications.

Head, B. (2010). Evidence-based policy: Principles and requirements in strengthening evidence-based policy in the Australian federation. *Roundtable Proceedings, Canberra: Productivity Commission*, 13.

Howes, M., McKenzie, M., Gleeson, B., Gray, R., Byrne, J. and Daniels, P. (2010). Adapting ecological modernisation to the Australian context. *Journal of Integrative Environmental Sciences*, 7(1), 1–17.

Huber, J. (1991). Ecological modernisation: Beyond scarcity and bureaucracy. *Technologie En Milieubeheer*, 167–183.

Hyman, R. (2001). *The strategic orientations of trade unionism*. London: Sage Publications.

ICC (International Chamber of Commerce). (2010). *Business, part of the solution*. Paris: International Chamber of Commerce.

ICC. (2012). *Green economy roadmap* (No. Doc 213-18-8). Paris: International Chamber of Commerce.

ICC. (2015a). *ICC call for climate action*. Paris: International Chamber of Commerce.

ICC. (2015b). ICC constitution (June 2015), preamble. Retrieved on 22 November 2016 from www.iccwbo.org/constitution/.

IISD. (2016). 175 parties sign Paris Agreement, 15 ratify agreement on earth day. *Sustainable Policy and Practice*, 22 April.

ILO. (2011a). Decent work agenda. Retrieved on 4 November 2016 from www.ilo.org/global/about-the-ilo/decent-work-agenda/lang--en/index.htm.

ILO. (2011b). *International Labour Organization submission to the UNFCCC on the Cancun adaptation framework on enhanced action and adaptation*. Geneva: International Labour Organization.

ILO. (2011c). *Skills and occupational needs in green building*. Geneva: International Labour Organization, Skills and Employability Department.

ILO. (2011d). *Skills and occupational needs in green building*. Geneva: International Labour Organization, Skills and Employability Department.

ILO. (2012). *Working towards sustainable development: Opportunities for decent work and social inclusion in a green economy*. Geneva: International Labour Office.

IOE (International Organisation of Employers). (2009). *Climate change: Information paper*. Geneva: International Organisation of Employers.

IPCC (Intergovernmental Panel on Climate Change). (2014). *IPCC WGII AR5 phase I report. climate change 2014: Impacts, adaptation, and vulnerability* (WGII AR5 Phase I Report). Tokyo, Japan: Intergovernmental Panel on Climate Change.

ITCILO. (2015). A guide on CSR and human rights – what does it mean for companies in supply chains? Brussels: International Training Centre of the International Labour Organization and BusinessEurope.

ITUC. (2011). What do we want in Cancun? A just transition now! Retrieved on 7 November 2016 from www.ituc-csi.org/what-do-we-want-in-cancun-just.html?lang=en.

ITUC. (2012a). Growing green and decent jobs. Retrieved on 8 November 2016 from www.ituc-csi.org/IMG/pdf/ituc_green_jobs_summary_en_final.pdf.

ITUC. (2012b). There will be no social justice without environmental protection. Retrieved on June 29 2012 from www.sustainlabour.org.

Janicke, M. (2008). Ecological modernisation: New perspectives. *Journal of Cleaner Production*, 16(5), 557–565.

Joey, A. (2015). The role of non-state actors in international relations. Retrieved on 7 November 2016 from www.academia.edu/5124220/The_Role_of_Non-state_Actors_in_International_Relations.

Jokinen, P. (2000). Europeanisation and ecological modernisation: Agri-environmental policy and practices in Finland. *Environmental Politics*, 9(1), 138–167.

Kelly, J. and Heery, E. (1994). *Working for the union*. London: Cambridge University Press.

Kjaer, P.F. (2015). *The evolution of intermediary institutions in Europe: From corporatism to governance*. Basingstoke: Palgrave Macmillan.

Kooiman, J. (2000). Societal governance: Levels, models, and orders of social-political interaction. In J. Pierre (ed.), *Debating Governance: Authority, Steering and Democracy*. Oxford: Oxford University Press, pp. 138–166.

Krauss, C. and Bradsher, K. (2015). Climate deal sends signal for industry to go green. *International New York Times*, 15 December, pp. 1–18.

Lammerts van Beuren, E.M. and Blom, E.M. (1997). *Hierarchical framework for the formulation of sustainable forest management standards*. Leiden, Netherlands: The Tropenbos Foundation.

Lange, H. and Garrelts, H. (2007). Risk management at the science-policy interface: Two contrasting cases in the field of flood protection in Germany. *Journal of Environmental Policy and Planning*, 9(3–4), 263–279.

Langhelle, O. (2000). Why ecological modernisation and sustainable development should not be conflated. *Journal of Environmental Policy and Planning*, 2(4), 303–322.

Laurent, E. (2010). *Environmental justice and environmental inequalities: A European perspective* (Issue 3/2010). Brussels: ETUI.

Levitt, R. and Solesbury, W. (2005). *Evidence-informed policy: What difference do outsiders in Whitehall make?* (Working Paper No. 23). London: ESRC UK Centre for Evidence Based Policy and Practice.

Lock, G. (2006). The big ask gets top-quality response. *Third Sector, London* (420), 14.

LSE Grantham. (2015). *The 2015 global climate legislation study: Climate change legislation in Kenya*. London: Grantham Research Institute on Climate Change and the Environment.

Macquarie. (2013). Employers' associations. Retrieved on 7 November 2016 from www.credoreference.com.ezproxy.bond.edu.au/entry/mdpo/employers_associations.

Mallet, V. (2016). Energy viability concerns: Confidence in India solar power plans start to fade. *Financial Times*, 25 April, p. 18.

Maraseni, T.N. and Cadman, T. (2015). A comparative analysis of global stakeholders' perceptions of the governance quality of the clean development mechanism (CDM) and reducing emissions from deforestation and forest degradation (REDD+). *International Journal of Environmental Studies*, 72(2), 288–304.

Marsden, T., Yu, L. and Flynn, A. (2011). Exploring ecological modernisation and urban-rural eco-developments in china: The case of Anji county. *The Town Planning Review*, 82(2), 195–225.

MEDEF. (2013). *Entreprises et biodiversité comprendre et agir*. Paris: Mouvement des Entreprises de France.

MEDEF. (2014). *Guide sur les initiatives RSE sectorielles première édition. Les fédérations professionnelles s'engagent pour la RSE*. Paris: Mouvement des Enterprises de France.

MEDEF. (2015a). Joint statement high-level business summit on energy and climate change, December 8 and 9, MEDEF headquarters, Paris. Paris: Mouvement des Enterprises de France.

MEDEF. (2015b). *Le MEDEF lance le manifeste des entreprises pour la conférence climat Paris 2015 (COP 21): Les entreprises proposent des solutions!* Paris: Mouvement des Enterprises de France.

MENR. (2015). *Kenya's national climate change action plan.* Nairobi: Kenya Ministry of Environment and Natural Resources.

Merriam-Webster. (2013). Theory definition. Retrieved on 7 November 2016 from www.merriam-webster.com/dictionary/theory.

MEWR. (2015). *Sustainable Singapore blueprint 2015*. Singapore: Ministry of the Environment and Water Resources.

Meyer, J., Boli, J., Thomas, G. and Ramirez, F. (1997). World society and the nation-state. *The American Journal of Sociology*, 103(1), 144–181.

Ministère de l'Écologie. (2010). *Green growth mobilisation plan.* Retrieved on 7 November 2016 from www.developpement-durable.gouv.fr/.

Ministry of Ecology, Sustainable Development and Energy. (2013). *Climate and energy efficiency policies. Summary of France's undertakings and results.* Paris: French Ministry of Ecology, Sustainable Development and Energy.

Ministry of Ecology, Sustainable Development and Energy. (2015). *The ecological transition towards sustainable development: A new strategy for 2015–2020.* Paris: French Ministry of Ecology, Sustainable Development and Energy.

Miranda, G. and Larcombe, G. (2012). *Enabling local green growth: Addressing climate change effects on employment and local development.* Paris: OECD Local Economic and Employment Development (LEED).

Mol, A. (2001). *Globalisation and environmental reform.* London: Massachusetts Institute of Technology.

Mol, A. (ed.) (2008). *Environmental reform in the information age.* New York: Cambridge University Press.

Mol, A. and Sonnenfeld, D. (2000). Ecological modernisation around the world: An introduction. *Environmental Politics*, 9(1), 3–16.

Mol, A., Sonnenfeld, D. and Spaargaren, G. (eds) (2009). *The ecological modernisation reader: Environmental reform in theory and practice* (1st edn). London: Routledge.

Moore, S.M. (2006). Theoretical framework. *ProQuest*. New York: Springer Publishing Company.

Morris, S. (2010). Fresh face, ACTU role makeover. *Australian Financial Review*, 5 July.

Nasiritousi, N. and Linnér, B.-O. (2016). Open or closed meetings? Explaining nonstate actor involvement in the international climate change negotiations. *International Environmental Agreements: Politics, Law and Economics*, 16(1), 127–144.

NTUC. (2016). About national trade union congress. Retrieved on 8 November 2016 from www.ntuc.org.sg/wps/portal/up2/home/aboutntuc.

OECD. (2010). *United Kingdom policies for a sustainable recovery*. Paris: Organisation for Economic Co-operation and Development.

OECD. (2011). *Green growth strategy synthesis report* (No. ECO/CPE/WP1 (2011)2/ADD1). Paris: Organisation for Economic Co-operation and Development, Economic Policy Committee.

OECD. (2013a). *Putting green growth at the heart of development*. Paris: Organisation for Economic Cooperation and Development.

OECD. (2013b). *What have we learned from the attempts to introduce green growth policies?* Paris: Organisation for Economic Co-operation and Development.

OECD. (2016). Greening jobs and skills. Retrieved on 7 November 2016 from www.oecd.org/employment/greeening/jobsandskills.htm.

Oxford Dictionary. (2015). Non-state actors. Retrieved on 7 November 2016 from www.oxforddictionaries.com/definition/english/non-state-actor.

Parsons, W. (2002). From muddling through to muddling up – evidence based policy making and the modernisation of British government. *Public Policy and Administration*, 17(3), 43–60.

Pellow, D., Schnaiberg, A. and Weinberg, A. (2000). Putting the ecological modernisation thesis to the test: The promises and performances of urban recycling. *Environmental Politics*, 9(1), 109–137.

Pfeifer, S. and Sullivan, R. (2008). Public policy, institutional investors and climate change: A UK case-study. *Climatic Change*, 89(3–4), 245–262.

Pianigiani, G. (2015). World mayors join Pope in push for climate pact. *International New York Times*, 22 July, 4.

Pierre, J. and Peters, B. Guy. (2000). *Governance, politics and the state*. London: Macmillan.

Plowman, D. (1978). Employer associations: Challenges and responses. *Journal of Industrial Relations*, 20, 237–262.

Porter, M.E. (1990). *The competitive advantage of nations*. London: Macmillan.

Porter, M.E. and van der Linde, C. (1995). Green and competitive: Ending the stalemate. *Harvard Business Review*, 73(5) (September–October), 120–134.

Press Trust of India. (2016). WTO ruling against Paris pact. *The Sunday Statesman, Kolkata*, 28 February, p. 8.

Rademackers, K. and van der Laar, J. (2014). *Assessing the implications of climate policy adaptation on employment in the EU – final report and annexes*. Brussels: European Commission.

Radford University. (2012). Environmental history timeline. Retrieved on 8 November 2016 from www.radford.edu/-wkovarik/envhist/4industrial.html.

Rajamani, L. (2012). The Durban platform for enhanced action and the future of the climate regime. *International and Comparative Law Quarterly*, 61(2), 501–518.

Rametsteiner, E., Pülzl, H., Alkan-Olsson, J. and Frederiksen, P. (2011). Sustainability indicator development – science or political negotiation? *Ecological Indicators*, 11(1), 61–70.

Republic of Singapore. (2012). *National climate change strategy 2012*. Singapore: Republic of Singapore, Office of the Prime Minister.

Rhodes, R.A.W. (1997). *Understanding governance: Policy networks, governance, reflexivity and accountability*. Buckingham: Open University Press.

Ryan, D. (2012). COP 18 wrap-up: Weak Doha outcome. Retrieved on 7 November 2016 from www.theclimategroup.org/what-we-do/news-and-blogs/cop-18-wrap-up.

Salomon, L. (2002). The new governance and the tools of public action: An introduction. In L. Salomon (ed.), *The tools of government: A guide to the new governance*. Oxford: Oxford University Press, pp. 1–47.

Scheer, A. and Hoppner, C. (2010). The public consultation to the UK climate change act 2008: A critical analysis. *Climate Policy*, 10, 261–276.

Schmidt, J. (2008). Why Europe leads on climate change. *Survival: Global Politics and Strategy*, 50(4), 83–96.

Scott, M. (2014). Climate change: Implications for employment. Brussels: European Trade Union Institute.

Simonis, U. (1989). Ecological modernisation of industrial society. 3 strategic elements. *International Social Science Journal*, 41(08), 347–361.

Singapore Compact. (2014). *Singapore compact for corporate social responsibility annual report 2014*. Singapore: Singapore Compact for Corporate Social Responsibility.

Smismans, S. (2008). The European social dialogue in the shadow of hierarchy. *Journal of Public Policy*, 28(1), 161–180.

Smith, A. (1996). Mad cows and ecstasy: Chance and choice in an evidence-based society. *Journal of the Royal Statistical Society*, 159(3), 367–383.

Sonnenfeld, D. (2000). Contradictions of ecological modernisation: Pulp and paper manufacturing in south-east Asia. *Environmental Politics*, 9(1), 232–256.

Sonnenfeld, D. and Mol, A. (2006). Environmental reform in Asia: Comparisons, challenges, next steps. *The Journal of Environment and Development*, 15(2), 112–137.

Spaargaren, G., Mol, A. and Sonnenfeld, D.A. (2009). Ecological modernisation: Assessment, critical debates and future directions. In A. Mol, D. Sonnenfeld and G. Spaargaren (eds), *The ecological modernisation reader: Environmental reform in theory and practice*. London: Routledge, pp. 501–520.

Stephens, T. (2016). Signing the Paris climate agreement is easy – what comes next for Australia will be hard. *The Weekend Conversation*, 22 April.

Stern, N. (2007). *The economics of climate change*. Cambridge: Cambridge University Press.

Strietska-Ilina, O., Hofman, C., Duran Haro, M. and Jeon, S. (2011). *Skills for green jobs: A global view synthesis report based on 21 country studies. Executive summary*. Geneva: International Labour Organization.

Syndex. (2011). *Initiatives involving social partners in Europe on climate change policies and employment: Summary report for the conference on 1 and 2 March 2011*. Brussels: European Commission.

Taylor, L., Rezai, A. and Foley, D. (2016). An integrated approach to climate change, income distribution, employment, and economic growth. *Ecological Economics*, 121, 196–205.

Thomas, J. (2013). RIO + 20: Toward a new green economy – or a green-washed old economy? Retrieved on 7 November 2016 from http://grist.org/climate-policy/2011-03-24-rio-20-toward-a-new-green-economy-or-a-green-washed-old-economy/.

Traxler, F. (2010). The long term development of organised business and its implications for corporatism: A cross-national comparison of membership, activities and governing capacities of business interest associations, 1980–2003. *European Journal of Political Research*, 49, 151–173.

TUC. (2009). *Unlocking green enterprise. A low-carbon strategy for the UK economy*. London: Touchstone Pamphlets.

TUC. (2010). *GreenWorks: TUC GreenWorkplaces project report 2008–10*. London: Trades Union Congress.

TUC. (2012a). Greener deals: Negotiating on environmental issues at work. Retrieved on 8 November 2016 from www.tuc.org.uk/extras/greener_deals.pdf.

TUC. (2012b). About the TUC. Retrieved on 8 November 2016 from www.tuc.org.uk/the_tuc/index.cfm?mins=2&minors=2&majorsubjectID=19.

TUC. (2015). GreenWorkplaces news, October 2015. Retrieved on 21 November 2016 from www.tuc.org.uk/workplace-issues/green-workplaces/green-workplaces-news/greenworkplaces-news-october-2015.

TUC. (2016a). Environment. activist resource. Retrieved on 7 November 2016 from www.tuc.org.uk/social-issues/environment.

TUC. (2016b). Green Workplaces. Retrieved on 7 November 2016 from www.tuc.org.uk/workplace-issues/green-workplaces.

UK CCC. (2016). Committee on climate change, about us. Retrieved on 8 November 2016 from www.theccc.org.uk/about/.

UK Government. (1999). *White Paper: Modernising government*. London: UK Government.

UK NAO. (2016). *Investigation into the department of energy and climate change loans to the green deal finance company* (No. HC 888). London: UK National Audit Office.

UK POST. (2002). Postnote: Ratifying Kyoto. UK Parliamentary Office of Science and Technology, April (176).

UN Global Compact. (2016). United Nations global compact. Retrieved on 8 November 2016 from www.unglobalcompact.org/what-is-gc/mission.

UN SDKP. (2016). Sustainable development knowledge platform: Major groups and other stakeholders. Retrieved on 21 November 2016 from https://sustainabledevelopment.un.org/mgos.

UN. (1992). *United Nations Framework Convention on Climate Change*. Retrieved on 8 November 2016 from http://unfccc.int/files/essential_background/background_publications_htmlpdf/application/pdf/conveng.pdf.

UNCED. (1992). *Rio declaration on Environment and development*. Rio de Janeiro: United Nations.

UNCSD. (2012a). The future we want. *United Nations Conference on Sustainable Development*, 1–53.

UNCSD. (2012b). UNCSD Rio + 20 major groups. Retrieved on 21 November 2016 from www.uncsd2012.org/majorgroups.html.

UNFCCC. (1997). *Kyoto Protocol*. Kyoto: United Nations Framework Convention on Climate Change.

UNFCCC. (2007). Bali Action Plan (1/Cp.13), Report of the UNFCCC Conference of the Parties (COP 13), UN Doc. No. FCCC/CP/2007/6/Add.1, at 3.

UNFCCC. (2010). *Climate change conference in Cancun delivers balanced package of decisions, restores faith in multilateral process*. Cancun: United Nations Framework Convention on Climate Change.

UNFCCC. (2011a). Parties and observers. Retrieved on 7 November 2016 from http://unfccc.int/parties_and_observers/items/2704.php.

UNFCCC. (2011b). *Durban platform for enhanced action*, Durban, South Africa: United Nations Framework Convention on Climate Change Conference of the Parties (COP 17).

UNFCCC. (2013). Principle of common but differentiated responsibilities. Retrieved on 7 November 2016 from http://unfccc.int/essential_background/convention/background/items/1355.php.

UNFCCC. (2015). *Paris agreement*. Paris: United Nations Framework Convention on Climate Change.

UNFCCC AWG LCA. (2009). Report of the Ad Hoc Working Group on long-term cooperative action under the convention, UN Doc. No. FCCC/AWGLCA/2009/17.

UNFCCC LPAA. (2015). Conference of Parties [COP]/Meeting of parties to the Kyoto Protocol [CMP] Peruvian Presidency, French Presidency, United Nations Framework Convention on Climate Change [UNFCCC] Secretariat. *The Lima Paris action agenda [LPAA]. Vision and approach.* Paris: Paris 2015 UN Climate Change Lima.

UNFCCC LPAA. (2016). Retrieved on 7 November 2016 from http://newsroom.unfccc.int/lpaa/about/#LPAA Presentation.

UNHLPF (United Nations High Level Political Forum). (2013). UN member states begin negotiations on HLPF. Retrieved on 7 November 2016 from http://uncsd.iisd.org/news/un-member-states-begin-negotiations-on-hlpf/.

University of Edinburgh. (2012). The London smog disaster of 1952. Retrieved on 8 November 2016 from www.portfolio.mvm.ed.ac.uk/studentwebs/session4/27/greatsmog52.htm.

van den Hove, S. (2000). Participatory approaches to environmental policy-making: The European commission climate policy process as a case study. *Ecological Economics*, 33, 457–472.

van Kersbergen, K. and F. van Waarden, F. (2004). Governance' as a bridge between disciplines: Cross-disciplinary inspiration regarding shifts in governance and problems of governability, accountability and legitimacy. *European Journal of Political Research*, 43(2), 143–171.

WCED (World Commission on Environment and Development). (1987). *Report of the world commission on environment and development: Our common future* (A/42/427). New York, NY: United Nations.

WEF. (2011). *A profitable and resource efficient future: Catalysing retrofit finance and investing in commercial real estate* (No. 031011). Geneva: World Economic Forum.

Weiss, D.G., Seyle, D.C. and Coolidge, K. (2013). The rise of non-state actors in global governance. Opportunities and limitations. *One Earth Future Discussion Paper*, 1–29.

WHO. (2011). *Guidelines on preventing early pregnancy and poor reproductive outcomes among adolescents in developing countries*. Geneva: World Health Organization.

Wilke, N. (2011). *Germany's climate policies towards a low carbon society: UNFCCC workshop on mitigation for developed countries*, 3 April 2011, Berlin: German Federal Ministry for the Environment, International Climate Policy Unit.

World Bank. (2013). Defining civil society. Retrieved on 7 November 2016 from http://web.worldbank.org/WBSITE/EXTERNAL/TOPICS/CSO/0.

Worldwatch Institute. (2008). *Green jobs: Towards decent work in a sustainable, low-carbon world.* Nairobi, Kenya: UNEP/ILO/IOE/ITUC.

WTO (World Trade Organization). (2012). *The outcomes of Rio + 20: A seminar hosted by the WTO*. Geneva: World Trade Organization.

WWF. (2003). *Power switch: Impacts of climate policy on the global power sector*. London: World Wide Fund for Nature.

WWF. (2011). *Summary report: EU climate policy tracker 2011*. Brussels: World Wide Fund for Nature.

York, R. and Rosa, E. (2003). Key challenges to ecological modernisation theory: Institutional efficacy, case study evidence, units of analysis and the pace of eco-efficiency. *Organisation Environment*, 16(3), 273–288.

Index